STRATEGIES *for* SUCCESS

MATH

Problem Solving

ategies for Success: Math Problem Solving, Grade 4 OT112 / 323NA ISBN-13: 978-1-60161-937-2
ver Design: Bill Smith Group **Cover Illustration:** Valeria Petrone/Morgan Gaynin

umph Learning® 136 Madison Avenue, 7th Floor, New York, NY 10016

ted in the United States of America. 10 9 8 7 6 5 4 3 2 1

Table of Contents

Problem-Solving Toolkit

How to Solve Word Problems

Have you ever tried to get somewhere without clear directions? How did you find a way to get there? You can use the same kind of thinking when solving word problems.

- ☐ **Read the Problem** Read carefully to be sure you understand the problem and what it is asking. Try to get a picture in your mind of what is going on and what is being asked.

- ☐ **Search for Information** Look through all the words and all the numbers to see what information is given. Study any charts, graphs, and pictures. Anything that might help you solve the problem is important.

- ☐ **Decide What to Do** Think about the problem. If you are not sure how to solve it right away, ask yourself if you have solved a problem like this before. Think about all the problem-solving strategies you know. Choose one that you think will work.

- ☐ **Use Your Ideas** Start to carry out your plan. Try your strategy. Think about what you are doing. Once in a while, ask yourself, *Am I on the right track?* If not, change what you are doing. There is always something else you can try.

- ☐ **Review Your Work** Keep thinking about the problem. Finding an answer does not mean you are done. You need to keep going until you are sure you solved the problem correctly.

You can use the Problem-Solving Checklist on page 7 to make sure you have followed these important steps.

Problem-Solving Checklist

☐ Read the Problem

- ☐ Read the problem all the way through to get an idea of what is happening.
- ☐ Use context clues to help you understand unfamiliar words.

Ask yourself

- ☐ How can I restate the problem in my own words?

☐ Search for Information

- ☐ Reread the problem carefully with a pencil in your hand. Circle the important numbers and math words.

Ask yourself

- ☐ What do I already know?
- ☐ What do I need to find out to answer the question the problem asks?
- ☐ Does the problem have any facts or information that are not needed?
- ☐ Does the problem have any hidden information?
- ☐ Have I solved a problem like this before? If so, what did I do?

☐ Decide What to Do

- ☐ Choose a strategy that you think can help you solve the problem.
- ☐ Choose the operations you will use.

Ask yourself

- ☐ How can I use the information I have to solve the problem?
- ☐ Will this problem take more than one step to solve?
- ☐ What steps will I use?

☐ Use Your Ideas

- ☐ Try the strategy you chose to solve the problem.
- ☐ Do the necessary steps.
- ☐ Write a complete statement of the answer.

Ask yourself

- ☐ Do I need any tools such as a ruler or graph paper?
- ☐ Would an estimate of the answer help?
- ☐ Is my strategy working?

☐ Review Your Work

- ☐ Reread the problem.
- ☐ Check your computations, diagrams, and units.

Ask yourself

- ☐ Is my answer reasonable? Does it make sense?
- ☐ Did I answer the questions the problem asks?

Use Logical Reasoning

Who took that muffin? You want to find out. So you begin to search for clues. You look at the muffin pan, the table, and the floor. If you think logically about the facts, you should be able to rule out possible answers.

The bit of blueberry left behind tells you that your brother did not take the muffin. He would have gone for the banana muffin. Your sister is in the clear. She left the house while the muffins were still baking. Who else could it be? Aha! That explains the crumbs on Sparky's food dish!

Sometimes you can solve a math problem just by thinking logically. You might not have to add, subtract, multiply, or divide at all.

Guess, Check, and Revise

You may have heard the old saying, *If at first you don't succeed, try again.* You can use this idea with math problems, too. Just guess the answer and then see if it is correct. If not, try again. Try to make your next guess a better one by looking closely at why your first guess was not correct.

Item	Price	Quantity	Total
card	$5 each		
pen	$3 each		
	Total	11	$39

Look at the torn receipt on the right. Someone bought some cards for $5 each and some pens for $3 each. This person bought 11 items in all and spent a total of $39.

Could the person have bought 1 card and 10 pens? The total cost would be $35. That is too low.

So guess again. Because your guess was low, try more cards and fewer pens.

You can keep changing your guesses, getting closer and closer, until you get the correct answer.

When you use this strategy, be sure to use the information from each guess and check to help you make your next guess.

Write an Equation

Mmmmm, bet you cannot wait for some fresh carrots and beans to grow in your garden. But you have to plant the seeds first! How many plants will you grow? That depends on the area of the garden.

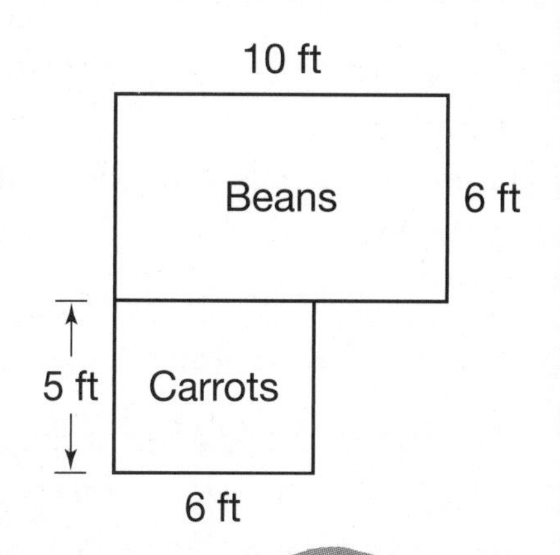

Since both sections of your garden are rectangles, you can use a special kind of equation called a formula.

Area of a rectangle = length × width

Area of bean section = 10 feet × 6 feet = 60 square feet
Area of carrot section = 6 feet × 5 feet = 30 square feet

The formula helps you find the total area of the garden.

You can use the same formula to find the area of any rectangle, no matter how big or how small.

Make a Table

Have you seen how bands in parades keep in step in rows and columns?

Like bands, information can be organized in rows and columns. A table organizes information in this way. A table can help you see how data are related. You can make and use tables to solve math problems.

You are saving money to buy a new bicycle that costs $89. You have $47 and you will save $6 each week. How long will it take you to save enough money to buy the bike?

Make a table to show your savings week by week.

Time	Start	1 week	2 weeks	3 weeks
Amount	$47	$53	$59	$65

A table makes it easy to keep track of information.

You can complete the table to solve the problem.

Problem-Solving Strategies

Work Backward

If you know the end and you know operations, you can work backward to get to the start.

Whew! You followed the directions and made it to your friend's new house. But how will you find your way back home? It is simple. Just follow the same directions, but work backward.

You can use this strategy of working backward to solve some math problems, too.

Your family bought a bag of apples. Your sister and her friends took 8 apples. Your brother took 3 apples. There are 4 apples left. How many apples were in the bag to begin with?

You can work backward by undoing the operations. To undo an operation, use the opposite operation. Start with 4, the end result. Add the 3 apples your brother took to get 7 apples. Then add the 8 apples your sister and her friends took to 7 to get 15 apples. So there were 15 apples in the bag to begin with.

Solve a Simpler Problem

Look at the two domino cards at the right. They each have 5 dots, but the dots are in different places.

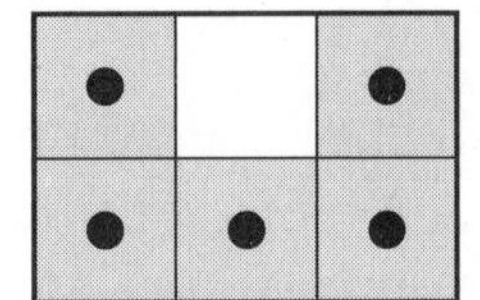

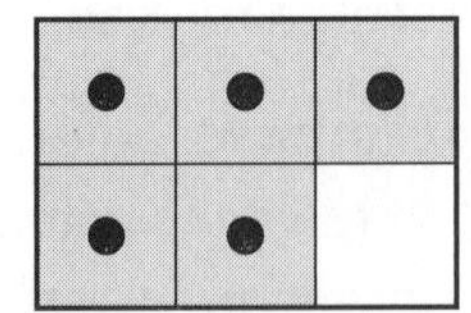

How many different ways can you arrange 5 dots on this kind of card?

You can make the problem simpler by thinking about the 1 empty box instead of 5 dots.

In how many different places can the empty box be?

The answer is the same, but it is a simpler problem to solve.

If a problem seems complicated, try to break it down into a simpler problem.

Draw a Diagram

How long do you think the longest train ever built was? As long as a soccer field? A mile? Guess again. It was over 4 miles long! It had about 500 cars!

How long would a train of 5 cars be if each car was 50 feet long and there was 5 feet between one car and the next?

You can draw a diagram that shows all the lengths.

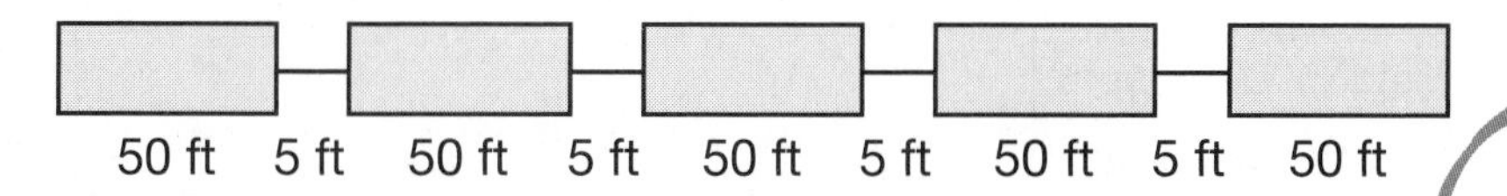

The diagram helps you see there are 5 cars but only 4 spaces between them.

You could say the diagram helps keep you and this train problem on track!

When you draw a diagram to solve a math problem

- keep it simple
- make sure the diagram shows the math in the problem
- use labels to help make the information clear

Look for a Pattern

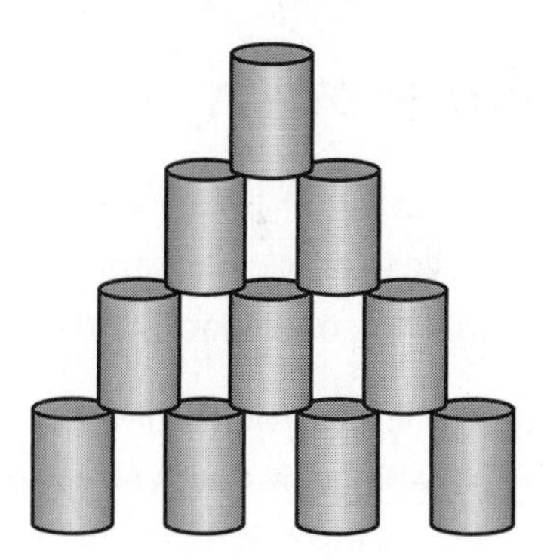

Watch your step! Do not knock over those cans!

Suppose you wanted to make the stack of cans 10 rows high. How many cans would you need?

You could find a pattern.

1 can in row 1

2 cans in row 2

3 cans in row 3

4 cans in row 4

There are $1 + 2 + 3 + 4$ cans in 4 rows.

You can use this pattern to figure out how many cans would be in 10 rows.

Be careful when you use patterns. What might look like a pattern at the beginning may not keep going.

Problem-Solving Strategies

Make a Graph

You know that pictures can help you to see what is going on in math problems. A graph is a kind of picture. It is a picture of data.

The table shows data for the five most popular types of books. It shows how many books of each type were bought. The graph shows the same data but in a different way.

Most Popular Books

Type of Book	Number Bought
History/Geography	244
Sports/Art/Music	3,064
Technology/Health	590
Science	769
Social Science	679

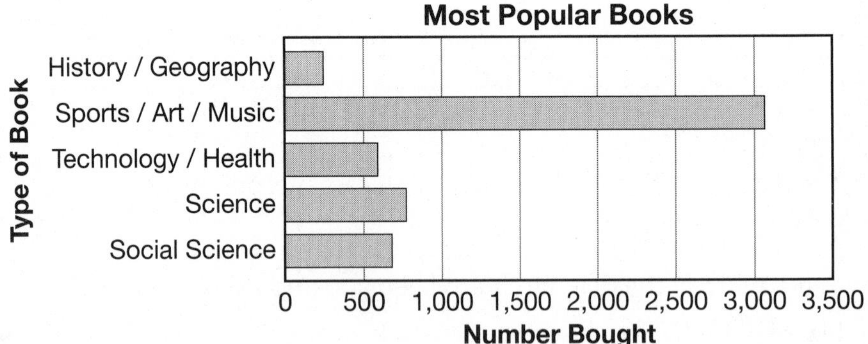

A graph can make it easier to see and compare the data.

Make an Organized List

How many combinations of two tacos are there? The tacos can be the same type or different types. List all the combinations to find out.

You can list all the combinations with beef, then all the combinations with chicken, and so on. You can use letters instead of words: *B* for Beef, *C* for Chicken, *V* for Vegetable, and *F* for Fish.

Tacos

Two Taco Plate $4.99

Mix and match any two types of tacos.

Beef Chicken Vegetable Fish

BB	BC	BV	BF
	CC	CV	BF
		VV	VF
			FF

Some combinations are not listed because they were listed already. For example, CB is not listed because CB is the same combination as BC.

How many different two taco combinations are there?

Making an organized list helps you make sure you do not leave anything out or list anything twice.

Check for Reasonable Answers

Wait a minute! You just bought a 99¢ box of crayons. You paid with a ten-dollar bill but you only got back a penny in change. Luckily, you know when something does not make sense.

Whenever you solve a problem, it is a good idea to look back to see if your answer is reasonable. Checking for a reasonable answer is a great way to find out if something is wrong. Then you can fix it.

Suppose you need to find the total number of fourth graders at Pine School. You add the number of students in each class and get 98.

You know that $4 \times 25 = 100$. Do you see why the total cannot be less than 100?

Pine School Fourth Grade

Class	Number of Students
Ms. Gomes	27
Mr. Lee	28
Mr. Johns	27
Ms. Simon	26
Total	

Decide If an Estimate or Exact Answer Is Needed

Friends for Life is finally showing at a theater near you. You want to go to the 1:25 P.M. show, but you have a soccer game later in the afternoon. You need to be out of the theater by 4 o'clock to make it to the game. Do you have time to see the movie?

ADMIT ONE

Cinema $8\frac{1}{2}$

Friends for Life 1 hr 52 min

1:25 P.M.

ADMIT ONE

You could try to figure out the exact time the movie will end. But do you really need to do all that work?

You can estimate instead. The movie is less than 2 hours long, and it starts at about 1:30. Do you see why it will end before 4:00?

Sometimes you do need an exact answer. Suppose you bring $15 to the movie and spend $10.75. You may want to know exactly how much money you should have left.

You know that an estimate can help you check an answer. But sometimes an estimate is all you need to find an answer.

Decide What Information Is Unnecessary for Solving

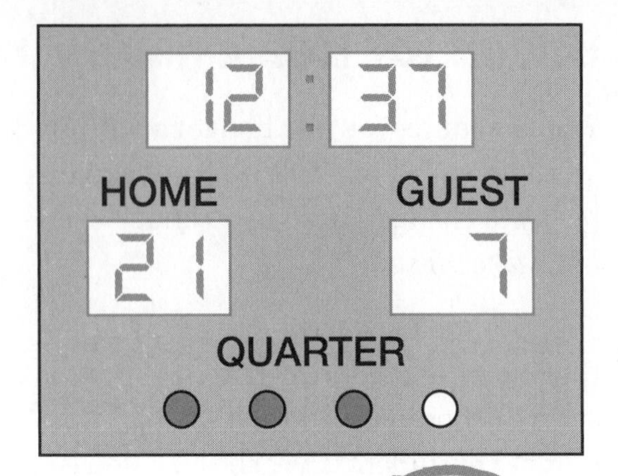

You just got to the game. People are yelling. Music is blasting. The scoreboard is full of numbers. Right now you do not need all that information. You just want to know the score.

You can zoom in on the information you need and ignore the rest. You can use this skill when you solve math problems, too.

Solving a math problem can be like looking at a scoreboard. It helps to know what you are looking for, so you can ignore information you do not need.

Find Hidden Information

You have come to the store to buy a board game. You see the one you want, but the price is not marked. So you look around and you see a sign with the information you need.

You might have to read a math problem more than once to find any hidden information.

Sometimes it may seem that a math problem does not give you all the information you need to solve it. Do not give up. Look around. Maybe the information is given in a picture or a table. Maybe the information you need is something you already know, such as 1 foot = 12 inches or that the month of May has 31 days.

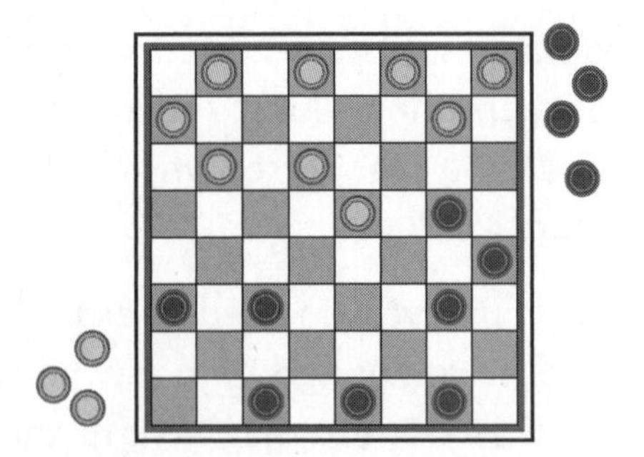

ALL BOARD GAMES: $10

Use Multiple Steps to Solve

Get ready for a riddle. How do you get to be a great dancer? One step at a time, of course.

It is no riddle that you may often need to complete more than one step to finish a math problem, too.

Here is an example that involves multiple math steps. How much would it cost for a group of 3 adults and 5 children to attend a show called *Dance, Dance, Dance*?

For a first step, you can find the cost for 3 adults.

$3 \times \$15 = \$45$

For a second step, you can find the cost for 5 children.

$5 \times \$8 = \$40$

To find the total cost, what will your next step be?

Is your final answer to a multiple-step problem correct? That often depends on whether your answer to each step was correct. So be sure to go over each step when checking your work.

Dance, Dance, Dance

Admission	
Adults	\$15
Children	\$8

Interpret Answers

Look at the picture on the right. Is it a vase? Or do you see two faces? Your eyes and brain can see the drawing in different ways.

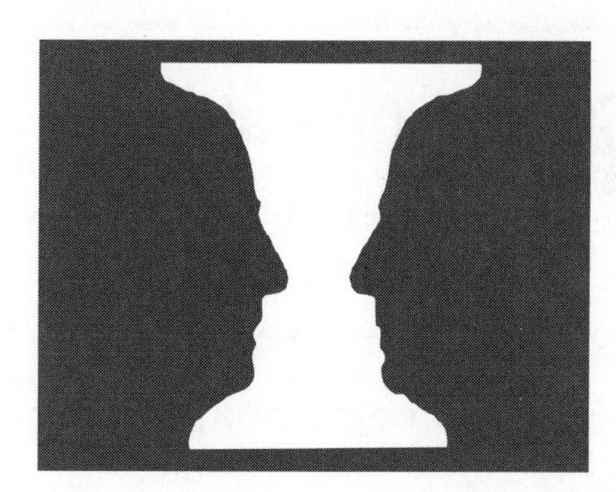

That is how solving some math problems can be. You get a number, but then you need to interpret it. You need to decide what the result means.

Think about the problem below.

You can pack up to 6 vases in a box. How many boxes do you need to pack 32 vases?

You can divide.

$$6\overline{)32} = 5 \text{ R2}$$

The number of boxes cannot be "5 remainder 2." Do you need 5 boxes or 6 boxes? You have to interpret 5 R2 so your answer makes sense.

There are different ways to interpret answers. Read the problem carefully so you know what is being asked.

Choose Strategies

A great thing about solving math problems is that you get to choose the way you think works best for you.

How do you want to get to the park? Do you want to skate there? Ride your bike? Maybe you would rather walk?

With math problems, there can be more than one way to get to the answer. Look at the problem below.

There are 32 teams entering a softball playoff. Each team will play one of the other teams. If a team loses a game, the team is out of the playoff. How many games will be played until there is only one team left?

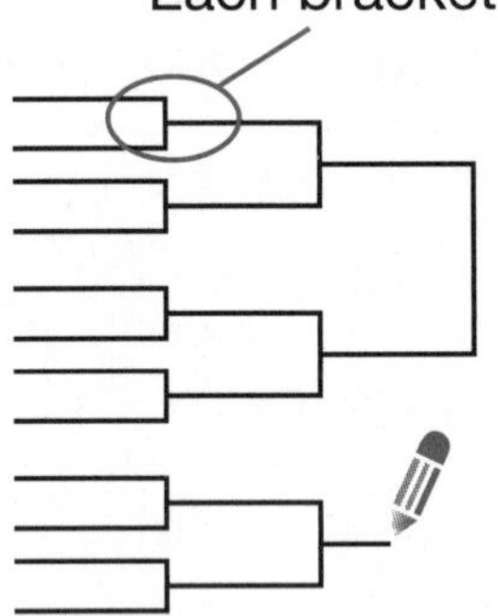

You might *draw a diagram* like the one on the right. You could keep drawing to show all 32 teams and then count all the games.

You might *look for a pattern*. How many games would be played if there are only 2 teams? Only 4 teams?

Or you might *use logical reasoning*. If one team gets eliminated in each game, how many games are played to eliminate all the teams except one?

Choose Operations

Get those skates on! It is opening day at the new ice rink. You and 5 friends can skate for $3 each. How much will it cost for the 6 of you to skate?

You can figure out the total cost in one step.

Find 6 × $3.

You can answer many questions with a single operation. Be sure to think about whether it makes sense to add, subtract, multiply, or divide.

When you solve a problem, always think first before calculating. You may be able to solve some problems without doing any operation at all.

Solve Two-Question Problems

When a math problem asks more than one question, take time to think about each one and answer it.

If you have a little brother or sister, you probably know what it is like to have questions coming at you like popcorn popping. *Where are you going? Can I come? What is that for?*

Math problems can also include more than one question. With math problems, you can focus on one question at a time. Once you have answered the first question, you can answer the next one.

The graph shows how fourth graders at Hill Elementary get to school. Which method do the most fourth graders use? Do more than half of the fourth graders get to school by bus?

Two questions require two answers. Can you figure out the answers to these questions?

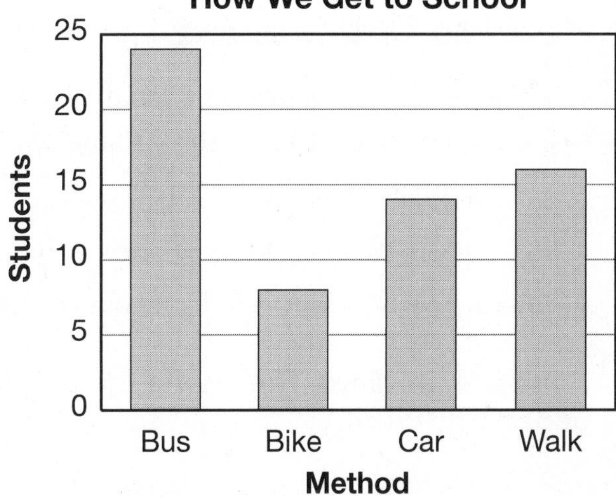

Formulate Questions

Why is the sky blue? Do dogs have dreams? How do our eyes work?
It is natural to wonder about things. It is natural to ask questions. The better you get at asking questions, the more you learn.

Look at this label from a box of cereal. You might wonder how many total grams of carbohydrates there are in each serving. What other questions might you ask?

Nutrition Facts	
Serving Size $\frac{2}{3}$ cup (32 g)	
Calories	130
Calories from Fat	25
Total Fat	3 g
Total Carbohydrate	25 g
Dietary Fiber	2 g
Sugars	6 g
Other Carbohydrates	17 g
Protein	2 g

Formulating questions about a math problem can help you understand it better.

How to Read Word Problems

Word problems usually do three things.

- They give you some information.
- They help you see how pieces of information are related.
- They give you a goal or ask a question.

You need to read a word problem carefully to understand it. Sometimes, reading again for a special purpose will help.

Read to Understand

When you read a word problem, you may see words or symbols that are new to you. Here are some things you can do.

- You can **look up** the word, fact, or symbol.
- You can **use context clues** to help you understand.
- You can use **base words** to figure out the meaning.

Read the problem. Then write the meaning of each underlined word.

The school driveway for buses is 10 yards longer than the driveway for cars. The car driveway is 90 feet long. Use the same units for your addends and then find the sum to get the length of the bus driveway.

1. I can look up the word units.

Meaning ______________________________

____

2. I can use context clues to figure out the meaning of addends.

Meaning ______________________________

____

3. I can use a base word to understand longer.

Meaning ______________________________

____

Sometimes, a word in a problem can have more than one meaning. Compare the math meaning of *yard* to its everyday meaning.

Everyday Meaning ______________________________

Math Meaning ______________________________

Look for Information

Read a problem once to be sure you understand what it is about. Then you can read it again to identify important numbers and words.

▶ A problem may give you all the details you need, and nothing else.

> You are selling tickets to a game. Each ticket costs $10.
> How much money will you take in by selling 20 tickets?

▶ A problem may have all the information you need and some extra details.

You do not need this detail to solve the problem.

> You are selling tickets to a game. You need to sell ~~30 tickets~~.
> If each ticket costs $10 and you have already sold 20 tickets, how much money have you taken in so far?

▶ A problem may have only some of the details you need. You must find a way to get the rest of the information.

You need to know how perimeter and length are related.

> A rectangular swimming pool has a perimeter of 600 feet.
> It is 200 feet long. How wide is it?

Some data you need may be in tables, graphs, or diagrams.

Read each problem. Study the information to decide how it can help you solve the problem. Then write your answer.

1. Mr. Polk wants to buy one of each small animal figure shown at the right. He pays with a $20 bill. How much change should he get?

$4.60 $3.20

$5.00

Information I can only get only from the price list: ______________________

____

2. You asked your friends which they like better: mac and cheese or tuna surprise. How many more prefer mac and cheese?

Which Do You Like Better?

	1	2	3	4
Mac and Cheese				
Tuna Surprise				

Data I can only get only from the graph: ______________________

____

Mark the Text

Marking information you need in the text as you read can help you organize your thinking.

- You can circle numbers, including numerals and words.
- You can cross out information you do not need.
- You can underline the question the problem asks.
- You can also underline something you do not understand or need to look up.

Sam's family is moving, so they rent a truck. It costs $19 for each of the first four hours. It costs $17 for each of the next four hours. It costs ~~$14 for each hour after that~~. How much will they pay to rent the truck for 7 hours?

Mark the text and tell why each mark is important.

On March 24, Ruth had $25 saved. That afternoon, she spent $12 on a shirt and $6 on a pair of flip-flops. After dinner, her aunt gave her $15, so Ruth went to a movie and spent $18. How much money did Ruth spend on March 24?

I underlined ______________________________

because ______________________________

I crossed out ______________________________

because ______________________________

I circled ______________________________

because ______________________________

Decide What to Do

You can look at a problem to be sure you know what question you need to answer.

► Sometimes a word problem clearly asks a question that you can answer by computing.

> You are selling tickets to a school play. You need to sell 30 tickets. Each ticket costs $10 and you have already sold 20 tickets. How much money have you taken in so far?

Explain What operations will you use? Explain how you decided.

___

___

► Sometimes you need to compute, but the result of your computation is not the answer to the problem.

> You are selling tickets to a school play. You need to sell 30 tickets. Each ticket costs $10 and you have taken in $220 so far. Have you sold all 30 tickets?

Determine How can you find out how much you will take in when you sell 30 tickets?

___

___

► Sometimes you do not need to compute at all. You can find the answer using a different method.

> The school is giving prizes to the 3 students who sold the most tickets. Name the top three sellers in order from least to greatest.

Ronnie	Bonnie	Connie	Donny	Lonny	Sonny	Honey
10	35	21	32	14	30	16

Analyze There is no question here. What do you need to do? What will your answer look like?

___

___

UNIT 1 Problem Solving Using Place Value, Addition, and Subtraction

Unit Theme:
Our World

Trips can take you far and wide. You might travel by bus or boat. You might even ride in a hot-air balloon. You will meet interesting people. Maybe you will attend a festival! It does not matter where you go. You can always celebrate the math in your world.

Math to Know

In this unit, you will use these math skills:

- Place value and number sense
- Add whole numbers
- Subtract whole numbers

Problem-Solving Strategies

- Use Logical Reasoning
- Guess, Check, and Revise
- Write an Equation
- Make a Table

Link to the Theme

Finish the story. Describe the parade.
Use both words and numbers.

Jin and his friends are at a dragon parade. The number of dragons in the parade is expected to set a record.

____

____

____

____

____

____

____

Use Math Language

Review Vocabulary

The list below shows vocabulary terms in this unit. Knowing the meaning of these terms will help you understand the problems.

difference	equation	pattern	rule
digit	expression	place value	sum

Vocabulary Activity Words in Context

Sometimes, context helps you figure out the meaning of words. Use terms from the list above to complete the following sentences.

1. The tens ________________ in the number 895 is 9.
2. To continue a ________________, you need to know the ______________.
3. In the number 512, the greatest ________________ is hundreds.
4. When you subtract one number from another number, you find the ________________.

Graphic Organizer Word Circle

Complete the graphic organizer.

- Find the two vocabulary terms from the list above that describe what is shown in the circle. Use the words to label the four parts.
- Cross out what does not belong in the circle.

$2 + 3 = 5$ ____________	$12 - 5 = 7$ ____________
$7 = 6 + 1$ ____________	$8 - 1$ ____________

Strategy Focus

Use Logical Reasoning

MATH FOCUS: Place Value

Learn About It

Read the Problem

544,270
1,295,178
12,604,767
36,961,664

Marta is doing a report on four states. The states are California, Hawaii, Pennsylvania, and Wyoming. She lists the number of people who lived in each state in 2009. But, she forgets to write the state names.

Marta remembers the facts below.

- California had the greatest number of people.
- Wyoming had the least number of people.
- About 10 times as many people lived in Pennsylvania as in Hawaii.

How many people lived in Hawaii in 2009?

Reread Ask yourself questions as you read.

- What is the problem about?

____

- What kind of information is given?

____

- What do you need to find?

____

Search for Information

Look for phrases that you will need to solve the problem.

Record What details will help you answer the question?

The states are California, Hawaii, ________________,
and ________________.

The numbers of people are 544,270; 1,295,178; ________________;
and ________________.

You can use this information to choose a problem-solving strategy.

Decide What to Do

You know the names of the four states. You know the numbers of people. You need to match a number with the right state.

Ask How can I use the clues to solve the problem?

- I can use the strategy *Use Logical Reasoning.*
- I can make a chart to organize the clues. I can use what I know to rule out possible answers.

Use Your Ideas

Step 1 Use the first clue: *California had the greatest number of people.* Put a ✔ in the box to show this. California cannot have any of the other *numbers of people.* So put an X in the empty boxes in that row. No other *state* can have 36,961,664 people. So put an X in the empty boxes in that column.

36,961,664 is the greatest number.

	544,270	1,295,178	12,604,767	36,961,664
California	X	X	X	✔
Hawaii				X
Pennsylvania				X
Wyoming				X

Step 2 Use the second clue: *Wyoming had the least number of people.* Use a ✔ to show this. Put an X in the empty boxes in the Wyoming row and in the 544,270 column.

Step 3 Use the third clue: *About 10 times as many people lived in Pennsylvania as in Hawaii.* Which of these two states has more people? ________________ Complete the chart.

So ________________ people lived in Hawaii in 2009.

Review Your Work

Check that you used the clues correctly.

State How does ruling out possible answers help you find the correct answer?

____

____

Try It

Solve the problem.

1. Paula has several fish tanks. Jimmy asks Paula how many fish she has altogether. She gives him these clues for the number.

   - The number is less than 200.
   - The number is more than 100.
   - The ones digit is 1 more than the hundreds digit.
   - The ones digit is 4 less than the tens digit.

   How many fish does Paula have?

Read the Problem and Search for Information

Read the clues. Identify clues that will help you get started.

Decide What to Do and Use Your Ideas

You can use the strategy *Use Logical Reasoning* to find the answer.

Ask Yourself

What will be the greatest place value in the number?

Step 1 Use the first and second clues. The number is less than 200 and more than 100.

So the hundreds digit must be ________ .

Step 2 Use the third clue. The ones digit is 1 more than the hundreds digit.

The hundreds digit is ________ . So the ones digit is ________ .

Step 3 Use the last clue. The ones digit is 4 less than the tens digit.

The ones digit is ________ . So the tens digit is ________ .

Paula has ________ fish.

Review Your Work

Reread the clues and make sure your answers match all of them.

Conclude Why should you start by finding the hundreds digit?

____

____

Apply Your Skills

Solve the problems.

2. Jay, Max, Tara, and Sam collect pop tabs from aluminum cans for a good cause. They collect 361; 282; 230; and 155 pop tabs.

- Max collects the least number of pop tabs.
- Jay collects about 200 more pop tabs than Max.
- Tara collects about 50 more pop tabs than Sam.

How many pop tabs does Tara collect?

	361	282	230	155
Jay				
Max				
Tara				
Sam				

Answer ______________________________

Explain How does using a chart help you solve the problem?

____

____

Hint Put a ✔ in a box to show the answer to the first clue.

Ask Yourself

Which number is about 50 more than another number?

3. All students at North School have ID numbers. Nicole forgot her number. Her teacher gives her clues to find the number.

- The number is greater than 700 and less than 800.
- The tens digit is 4 less than the hundreds digit.
- The sum of the tens digit and the ones digit is 8.

What is Nicole's ID number?

What is the greatest place value of the number? ______________

The hundreds digit must be ________ .

Answer ______________________________

Identify Which clues did you start with first and why?

____

____

Hint Use the clues to find each digit in the number.

Ask Yourself

How many digits will the number have?

Hint Find the hundreds digit and the thousands digits first.

4 Dee and her family drive to the mall. The parking lot sign shows the number of cars that can fit in the lot. Dee asks her brother Ben to guess the number. She gives him clues. The digits in the number are 2, 6, 7, and 9. The hundreds digit is 5 more than the thousands digit. The ones digit is even. What is the number?

Ask Yourself

Which number is 5 more than one of the other numbers?

The hundreds digit is ________ .

________ , ________ , ________ , ________

Answer ______________________________

Summarize How did the clue that the hundreds digit is 5 more than the thousands digit help you identify the hundreds digit?

____

____

Hint Read through all the clues to decide which one to use first.

5 Judy and Bo are studying countries with populations of less than 1 million people. They round the population numbers to the nearest thousand. Bo wants Judy to guess the population of the Solomon Islands. He gives her these clues.

- The hundreds digit is 5 more than the tens digit.
- The ones and tens digits are the same.
- The sum of the ones digit and the tens digit is 8.
- The thousands and ten thousands digits are both 2.
- The hundred thousands digit is 3 more than the thousands digit.

What is the population of the Solomon Islands *rounded to the nearest thousand*?

Ask Yourself

Which digit should I look at to round a number to the nearest thousand?

The ten thousands digit is ________ .

The thousands digit is ________ .

Answer ______________________________

____

Sequence What steps did you use to find the correct answer?

____

____

On Your Own

Solve the problems. Show your work.

6. Dina wants to guess the number of people at the high school football game. She has the following clues.
   - All three of the digits are odd.
   - The sum of the ones digit and the tens digit is 2.
   - The sum of all the digits is 5.

   How many people are at the game?

Answer ______________________________

Evaluate Why do you need the first clue to solve the problem?

____

____

7. The estimated numbers of people living in Evansville, Hanson, Bingham, and Clinton are listed in the table. Clinton has the greatest number of people. The number of people in Bingham is greater than the number of people in Hanson. If you round the populations of Hanson and Bingham to the nearest thousand, the numbers would be the same. How many people live in Evansville?

Estimated Population
14,989
13,824
14,046
15,452

Answer ______________________________

Justify Do you need to round all of the numbers? Explain.

____

____

Create

Look back at Problem 4. Think of a different number of cars that can fit in the parking lot. Write and solve a problem with a new set of clues.

Strategy Focus

Guess, Check, and Revise

MATH FOCUS: Addition and Subtraction

Learn About It

Read the Problem

On Monday, 345 people visited Victoria Falls. There were 215 fewer visitors in the morning than in the afternoon. How many people visited in the morning?

Reread Think about these questions as you read the problem.

- What is the problem about?

____

- Were there more visitors in the morning or more in the afternoon?

____

- What are you asked to find?

____

Mark the Text

Search for Information

Read the problem again. Circle the numbers and words you need to answer the question.

Record Write what you know about the problem.

There were ________ visitors in total.

There were 215 fewer visitors in the ____________ than in the ____________.

This information will help you choose a problem-solving strategy.

Decide What to Do

You know how the number of visitors in the morning compares to the number of visitors in the afternoon.

Ask How can I find the number of visitors in the morning?

- I can use the strategy *Guess, Check, and Revise.*
- First, I can guess the number of visitors in the afternoon. Then I subtract to find the number of visitors in the morning. I can check if my guess is correct by finding the sum of those numbers.

You can guess the numbers you do not know. Then you can change your numbers if you need to.

Use Your Ideas

Step 1 Try 250 for the number of afternoon visitors. Remember that the total number of visitors must be 345. Make a table to keep track of your guesses.

285 is too low. So there must have been *more* than 250 visitors in the afternoon.

Step 2 Try 300 for the number of visitors in the afternoon.

385 is too high. So there must have been *fewer* than 300 visitors in the afternoon.

Step 3 When you tried 300, the sum was too high. But it was closer to 345 than when you tried 250. Try 280 for the number of visitors in the afternoon.

Number in Afternoon	Number in Morning	Total Number of Visitors
250	$250 - 215 = 35$	$250 + 35 = 285$
300	$300 - 215 = 85$	$300 + 85 = 385$
280	$280 - 215 = \square$	$280 + \square = \square$

So there were ________ visitors in the morning.

Review Your Work

Check that you answered the question in the problem.

Identify Suppose your first guess was 290. How would you change your next guess?

___

___

Try It

Solve the problem.

1. Niagara Falls is made up of two waterfalls, Rainbow Falls and Horseshoe Falls. Together, they are about 923 meters wide. Horseshoe Falls is about 417 meters wider than Rainbow Falls. About how wide is each waterfall?

Read the Problem and Search for Information

Identify what you need to find. Reread and mark the numbers and words that will help you answer the question.

Decide What to Do and Use Your Ideas

You can use the strategy *Guess, Check, and Revise.*

Ask Yourself

How can I choose a first guess?

Step 1 Think of two numbers with a sum of about 900. The difference of the two numbers should be about 400. Try 650 meters for the width of Horseshoe Falls.

Width of Horseshoe Falls in Meters	Width of Rainbow Falls in Meters	Total Width in Meters
650	650 − 417 = 233	650 + 233 = 883
670	670 − 417 = ☐	670 + ☐ = ☐

883 is too low.

Step 2 The first guess is 40 less than the actual total width. Try a greater width for Horseshoe Falls. Try 670 meters.

So Horseshoe Falls is about ________ meters wide.

Rainbow Falls is about ________ meters wide.

Review Your Work

Check that your answers match the problem.

Contrast Start by guessing that the width of Rainbow Falls is 260 meters. How would you find the width of Horseshoe Falls?

____

____

Apply Your Skills

Solve the problems.

(2) At the Great Barrier Reef in Australia, a visitor can take a helicopter tour followed by a boat tour. Together, these two tours last 12 hours. The boat tour lasts 8 hours more than the helicopter tour. Both tours start from the same place. Julia wants to take the helicopter tour at 9:00 A.M. and the boat tour at noon. Will she be able to do both? Explain.

Hint You need to find out if the helicopter tour will end before the boat tour begins.

Helicopter Tour Hours	Boat Tour Hours	Total Hours
1 hour	1 + 8 = ☐	1 + ☐ = ☐
2 hours	☐ + ☐ = ☐	☐ + ☐ = ☐

Ask Yourself

What is a good first guess for the length of the helicopter tour?

Answer ______________________________

Describe Suppose your first guess was 3 hours. How could you use that guess to help you make your next guess?

____

____

(3) An ocean tour in Hawaii involves traveling to a coral reef and back. It takes the same amount of time to travel each way. The tour also spends 45 minutes at the reef. The whole tour takes about 1 hour 35 minutes. If you begin to travel to the reef at 2:00 P.M., what time will you get there?

Hint Find the number of minutes in 1 hour 35 minutes.

Minutes to Get to the Reef	Minutes to Get Back	Total Travel in Minutes	Minutes Spent at the Reef	Total Tour Time
20		20 + ☐ = 40		40 + ☐ = ☐
25		☐ + ☐ = 50		☐ + ☐ = ☐

Ask Yourself

Does the total tour time you found equal 1 hour 35 minutes?

Answer ______________________________

Explain How did you find the number of minutes for the whole tour?

____

____

Hint You may not need all of the information given in the problem.

Ask Yourself

If I guess the number of children, do I need to add or subtract to find the number of adults?

④ Tickets to the Space Needle are $17 for adults and $9 for children. Dora visits with her class. There are 45 fewer adults than children. There are 75 people in all. How many children and how many adults go to the Space Needle?

Number of Children	Number of Adults	Total Number
55		

Answer ______________________

Determine Did you need the information about the ticket prices to solve the problem? Why or why not?

____

____

Hint First, find the total amount Robert spent the first day.

Ask Yourself

How much did Robert spend on lift tickets?

⑤ Robert went on a 2-day ski trip in Colorado. He spent $300 during the trip. Robert spent $40 less on the second day than on the first day. He spent $50 on a lift ticket each day. How much did he spend on things besides the lift ticket on the first day?

Spent on First Day	Spent on Second Day	Spent on Lift Tickets	Total Spent

Answer ______________________

Conclude How did you decide on your first guess?

____

____

On Your Own

Solve the problems. Show your work.

(6) Matt built a model of the Great Pyramid in Egypt. Sue also built a model of the pyramid. It is 6 inches taller than Matt's. The combined height of the two models is 18 inches. How tall is Matt's model?

Answer ______________________________

Diagram Describe a diagram you could make to solve this problem.

____

____

(7) Meteor Crater is in Arizona. It was formed when a meteorite crashed into Earth. One day, 130 people visited the crater during the first two hours it was open. In the first hour, 80 fewer people came than in the second hour. How many people came in the first hour?

Answer ______________________________

Relate Suppose 80 *more* people arrived in the first hour than in the second hour. How would that change the way you solve this problem?

____

____

Create

Look back at Problem 3. Change the total tour time. Write and solve a new problem.

Strategy Focus

Write an Equation

MATH FOCUS: Expressions and Equations

Learn About It

Read the Problem

On Monday, Mrs. Chan drove 55 miles in the morning and 20 miles in the afternoon. She will drive 10 fewer miles on Tuesday than on Monday. She has enough gas left to drive 60 miles. Does she have enough gas to drive for Tuesday's trip?

Reread Think about these questions as you read.

- What is the problem about?

____

- What kind of information is given?

____

- What am I asked to find?

____

Mark the Text

Search for Information

As you read the problem again, think about the details that are given.

Record What numbers will help you solve the problem?

Mrs. Chan drove ________ miles on Monday morning.

Mrs. Chan drove ________ miles on Monday afternoon.

Mrs. Chan plans to drive ________ fewer miles on Tuesday than on Monday.

Mrs. Chan has enough gas to drive ________ miles.

Use this information to decide how to solve the problem.

Decide What to Do

You know how the number of miles Mrs. Chan drove on Monday is related to the number of miles she will drive on Tuesday. You know how many miles she can drive with the gas she has left.

Ask How can I use what I know to solve the problem?

- I can use the strategy *Write an Equation* to find the number of miles Mrs. Chan drove on Monday.
- Next, I can write another equation to find the number of miles Mrs. Chan will drive on Tuesday.

You need to find the total number of miles Mrs. Chan drove on Monday before you can find the number of miles she can drive on Tuesday.

Use Your Ideas

Step 1 Write an equation to find the total number of miles Mrs. Chan drove on Monday.

Morning Miles + Afternoon Miles = Miles Driven on Monday

55 + ________ = ________

Step 2 Find how far Mrs. Chan will drive on Tuesday.

Miles Driven on Monday	−	Difference in Miles on Tuesday	=	Miles to Drive on Tuesday
75	−	________	=	________

Step 3 Compare that answer to 60.

________ ○ ________

Does Mrs. Chan have enough gas to drive for Tuesday's trip? ______

Mrs. Chan will drive ________ miles on Tuesday.

She has enough gas to drive ________ miles.

Review Your Work

Be sure that you answered the question.

Describe How does writing an expression for the total distance Mrs. Chan drove on Monday help you solve the problem?

____

____

Try It

Solve the problem.

① Three friends left at noon on a 40-minute ride to the top of a mountain. They stayed at the top for a half hour. The ride back down took 45 minutes. At what time did they get back?

Read the Problem and Search for Information

Identify the information you will use. Reread and underline the question.

Decide What to Do and Use Your Ideas

You can use the strategy *Write an Equation* to find the total number of minutes the friends spent on the ride.

Step 1 Write an equation for the total time of the trip. Write the numbers for the times you know. Then solve the equation.

Time to Top + Time at Top + Time Back = Total Time

40 + 30 + 45 = ________

I know that there are 60 minutes in 1 hour. What operation can I use to write 115 minutes as hours and minutes?

Step 2 Rewrite the total time as hours and minutes.

115 minutes = 1 hour ________ minutes

Step 3 Find the time the friends get back from their ride.

Count on 1 hour 55 minutes from noon.

Noon + (1 hour 55 minutes) = ____________

So the friends got back from the ride at ____________.

Review Your Work

Check that the question asked you to find a time.

Conclude You first wrote the equation in words. How did that help you solve the problem?

____

____

Apply Your Skills

Solve the problems.

② Jen is at the beach. She has $30. She spends $5 on lunch. Then she decides to rent a bike. It costs $13 an hour to rent a bike. Does she have enough money to rent a bike for 2 hours? If so, how much money will she have left? If not, how much more money does she need?

> **Hint** First, find how much money Jen has after buying lunch.

> **Ask Yourself** What equation shows the cost of renting a bike for 2 hours?

Starting Amount − Cost of Lunch = Money Left After Lunch

________ − ________ = ________

Cost of renting a bike for 2 hours: ________________

Answer ________________________________

Identify How did you decide if Jen had enough money to rent a bike for 2 hours?

____

____

③ Mr. Washburn took a subway from home to a museum. The ride was 35 minutes to the museum and 40 minutes to get home. He spent the same amount of time at the museum as he did riding to and from the museum. How long did he spend at the museum?

> **Hint** The problem tells you that the time spent at the museum equals the time spent riding the subway both ways.

> **Ask Yourself** How can I use words to write an equation?

Time to Museum + Time from Museum = ________________

________ + ________ = ________

Answer ________________________________

Explain How is writing an equation like telling a story?

____

____

Ask Yourself

How can I describe this situation with an equation?

Hint You can write two equations to solve the problem.

④ The Parker family is taking a rowboat trip. They need to rent rowboats. The rowboats will cost them $100. They will also rent tents for a total of $55. The family has $450 to spend for the trip. How much will they have left after renting the rowboats and tents?

Cost of Rowboats + Cost of Tents = Total Cost

________ + ________ = ________

$450 − Total Cost of Rowboat and Tents = Amount Left

Answer ____

Determine Which words in the problem tell you that you will need to subtract to answer the question?

____

____

Hint A *round trip* means going and returning. A *one-way trip* means just going, or just returning.

Ask Yourself

How much will 2 one-way tickets cost?

⑤ Ms. Martin wants to take the Eurostar train from Paris to London. She wants to return to Paris the next day. A round-trip ticket costs about 300 euros. A one-way ticket costs about 196 euros. How much will Ms. Martin save if she buys a round-trip ticket instead of 2 one-way tickets?

Cost of 2 One-way Tickets − Cost of Round-trip Ticket = Amount Saved

Answer ____

Verify What other equation did you write to help you solve the problem?

____

____

On Your Own

Solve the problems. Show your work.

6. A bus has 60 seats for passengers. There are three stops on the way to the baseball game. At the first stop, 35 people get on the bus. At the second stop, 18 people get on. Are there enough seats for 10 people to get on at the third stop?

Answer ______________________________

Consider Explain how you could have solved the problem in another way.

____

____

7. Rico takes the 8:00 A.M. train into the city. He gets to the science museum at 9:00 A.M. Rico spends 2 hours at the museum. Then he spends one hour eating lunch. After lunch, Rico spends 2 and a half more hours at the museum. It will take Rico 15 minutes to get to the train station. Does Rico have enough time to catch the 3:00 P.M. train home?

Answer ______________________________

Analyze What information is given that is not needed to solve the problem?

____

____

Create

Look back at Problem 6. Change the number of people who get on and off at each stop. Write and solve a problem using the new numbers.

Strategy Focus

Make a Table

MATH FOCUS: Number Patterns

Learn About It

Read the Problem

Marc arrives at a kite flying festival at 8:45 A.M. The festival begins at 9:00 A.M. At the start of the festival, there are 36 kites in the sky. Every hour on the hour, 45 more kites are added to the kites already in the sky. Marc will stay until there are at least 200 kites in the sky. What is the earliest time he will leave the festival?

Reread Ask yourself these questions as you read.

- What is the problem about?

____

- What different kinds of numbers are given?

____

- What do I need to find?

____

Mark the Text

Search for Information

Read the problem again. Cross out numbers that you do not need to solve the problem.

Record What information will help you decide how to solve the problem?

The number of kites in the sky at 9:00 A.M. is ________ .

The number of kites added every hour is ________ .

You can use this information to choose a problem-solving strategy.

Decide What to Do

You know the number of kites in the sky at 9:00 A.M. and the number of kites added every hour on the hour.

Ask How can I find when there are at least 200 kites in the sky?

- I can use the strategy *Make a Table* to relate the time to the number of kites.
- I can use a rule to find the number of kites in the sky at each hour. Then I can find the earliest time when at least 200 kites are in the sky.

Use Your Ideas

Step 1 Use the rule to complete the table. Add 45 to find the number of kites at 10:00 A.M. Record the result in the table.

Time	9:00 A.M.	10:00 A.M.	11:00 A.M.		
Number of Kites in the Sky	36				

+ 45 + 45 + 45 + 45

Step 2 Compare the number of kites in the sky at 10:00 A.M. to 200 kites.

________ is less than 200.

Step 3 Complete the table. Record the number of kites in the sky at each hour. Stop when a number in your pattern is at least 200.

So the earliest Marc can leave is ____________.

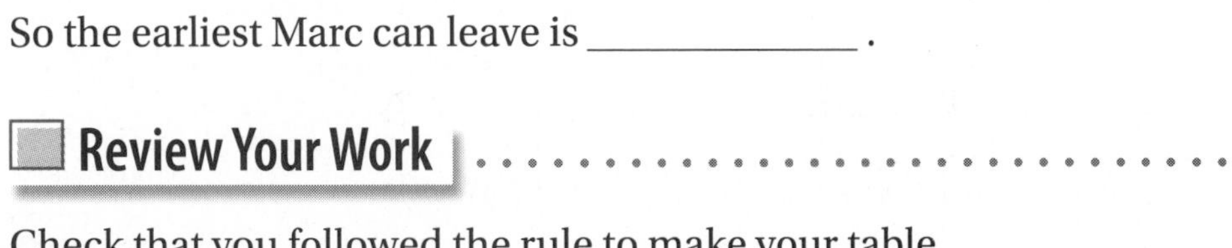

Review Your Work

Check that you followed the rule to make your table.

Conclude Why is making a table helpful in solving this problem?

____

____

Try It

Solve the problem.

1. In Chinatown, people hang strings of lanterns over streets to celebrate the Lantern Festival. Each lantern is hung on a string with a hook. The number of lanterns on each string follows the repeating pattern 14, 11, 14, 11... How many hooks are needed to hang all the lanterns on 7 strings?

Read the Problem and Search for Information

Retell the problem in your own words. Decide how you can find the total number of hooks needed to hang the lanterns.

Decide What to Do and Use Your Ideas

You can use the strategy *Make a Table.*

Ask Yourself

Why does the table stop at 7?

Step 1 Start with the number of lanterns on the first string.

Use this rule: add 11, add ________ .

Step 2 Use the rule to complete the table.

Number of Strings	1	2	3	4	5	6	7
Total Number of Lanterns	14						

+ 11 + 14 + ___ + ___ + ___ + ___

Step 3 There are a total of ________ lanterns on the 7 strings.

So ________ hooks are needed.

Review Your Work

Did you switch between adding 11 and 14?

Explain Suppose there are 8 strings of lanterns. How would you use the table to find the total number of hooks?

____

____

Apply Your Skills

Solve the problems.

2 Mr. Carey will bake apple pies for the Harvest Festival. The recipe shows the number of apples needed to make 2 pies. Mr. Carey has 23 apples. How many more apples does he need to make 12 pies?

Apple Pie Recipe
(makes 2 pies)

- 17 apples
- 1 cup sugar
- 4 tablespoons flour
- 2 teaspoons cinnamon
- 2 tablespoons butter
- 2 teaspoons lemon juice
- 2 pastry crusts

Hint The question asks how many more apples Mr. Carey needs, not how many apples are needed for 12 pies.

Ask Yourself

Should I add or subtract to find how many more apples Mr. Carey needs?

Number of Pies	2	4	6	8	10	12
Number of Apples	17	34				

Answer ______________________________

Summarize Which rules did you use to solve this problem?

____

____

3 One hundred ninety-eight chairs are set up in rows for a concert. There are 11 chairs in each row. How many rows of chairs should be taken away so that about 130 chairs are left for the next concert?

Ask Yourself

Does the question say there have to be exactly 130 chairs left?

Number of Rows Taken Away	0	1	2	3	4	5	6
Number of Chairs Left	198	187					

Hint Think about what rule to use to find how many chairs are left when a row is taken away.

Answer ______________________________

Predict Look at the table. Without subtracting, how could you find the number of chairs left if 7 rows are taken away?

____

____

Hint The question asks for the number of dancers, not the number of pairs.

Ask Yourself

If I know the number of pairs, how can I find the total number?

④ A large dance festival was held in Spain this year. In one performance, dancers made a pattern on the stage. In the first row, there were 2 pairs of dancers. Each row had 2 more pairs of dancers than the row in front of it. How many dancers were in the sixth row?

Row	1	2	3	4	5	6
Pairs of Dancers	2					
Number of Dancers						

Answer ______________________

Apply What other question can you answer about the number of dancers?

____

____

Hint Think about how you will label the rows.

Ask Yourself

If I decide to use the rule *add 28*, what number should I start with? If I decide to use the rule *subtract 28*, what number should I start with?

⑤ Mrs. Santos is decorating costumes. She has 148 roses. The table shows how many roses she needs for each type of costume. She decorates one crown. How many dresses can she decorate with the remaining roses? How many roses will she have left over?

Costume Types	Number of Roses Needed
Cape	32
Crown	5
Dress	28
Shirt	0

Answer ______________________

Analyze Leo uses the rule *subtract 28* to make his table. Julia uses the rule *add 28*. How can they both be correct?

____

____

On Your Own

Solve the problems. Show your work.

6. Hanna and her grandmother go for a ride in a hot air balloon. The balloon reaches a height of 1,000 feet in 20 minutes. Then the balloon goes down 180 feet every 3 minutes. How long from the start of the ride did it take to reach 100 feet on the way down?

Answer ______________________________

Contrast What two different rules did you use to find the answer? (Hint: One involves time; the other involves the balloon's height.)

____

____

7. Rosa is playing a basketball game at a street fair. Every time Rosa makes a basket, she wins tickets. For the first basket, she wins 1 ticket. For 2 baskets in a row, she wins 2 more tickets. For 3 baskets in a row, she wins 3 more tickets. Assume Rosa never misses a basket. How many baskets in a row does she need to make to win the piñata? How many tickets will she have left after paying for the piñata?

Answer ______________________________

Evaluate Raul thought that Rosa needed to make 40 baskets in a row to win the piñata. What was the error in Raul's thinking?

____

____

Look at Problem 2. Change the number of apples in the recipe. Then change the number of apples that Mr. Carey has. Write and solve a problem using a different number of pies.

UNIT 1 Review What You Learned

In this unit, you worked with four problem-solving strategies. You can often use more than one strategy to solve a problem. So if a strategy does not seem to be working, try a different one.

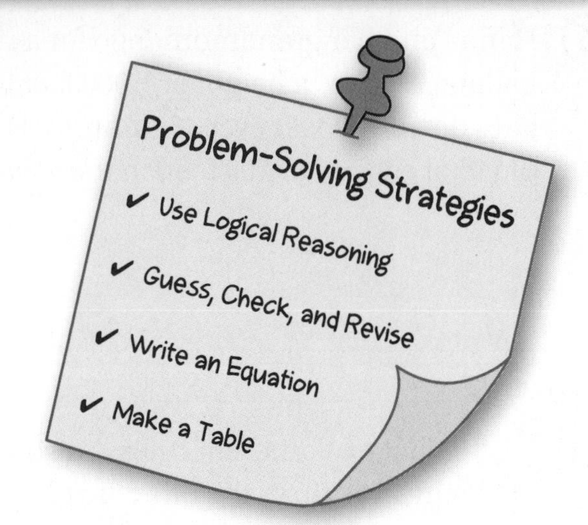

Solve each problem. Show your work. Record the strategy you use.

1. Fifteen people got on an elevator on the ground floor. On the fourth floor, 4 people got off. On the fifth floor, 5 people got on. How many people were on the elevator when it left the fifth floor?

Answer ______________________________

Strategy ______________________________

2. The total waiting time for two families to climb the Eiffel Tower was 55 minutes. One family waited 15 minutes longer than the other family. How long did each family wait?

Answer ______________________________

Strategy ______________________________

3. One concert ticket costs $10.00. Two tickets cost $9.50 each. Three tickets cost $9.00 each. If the pattern continues, how many tickets would you need to buy for each ticket to cost $7.00?

Answer ______________________

Strategy ______________________

4. There were 264 people who went to a play. Of those people, 136 arrived early and 73 arrived on time. The rest of the people arrived late. How many people arrived late?

Answer ______________________

Strategy ______________________

5. Mae is thinking of a 3-digit number. She gives these clues.

- The number is between 200 and 250.
- The sum of the ones digit and the tens digit is 6.
- The number is odd.
- The tens digit is less than the hundreds digit.

What is Mae's number?

Answer ______________________

Strategy ______________________

Explain how you used the clues to solve the problem.

Solve each problem. Show your work. Record the strategy you use.

6. Forty-six people are on a double-decker tour bus. There are 12 more people on the upper deck than on the lower deck. How many people are on each deck?

Answer ______________________________

Strategy ______________________________

7. Sixteen people walk in the park. If 5 more people are walking than skateboarding, how many people are skateboarding?

Answer ______________________________

Strategy ______________________________

8. One hundred people are waiting to take an elevator to the top of a tower. There are 30 more adults than children waiting for the elevator. How many adults are waiting for the elevator?

Answer ______________________________

Strategy ______________________________

Explain how you found your answer.

9. Jake saves money to take on a trip. He saves 1 dime on Monday. He saves 2 dimes on Tuesday. Each day he saves double the amount he saved the day before. How much money has Jake saved in all by Friday?

Answer ______________________

Strategy ______________________

10. Lilo is holding a card with a secret number. The number has 4 digits. The thousands digit equals the number of digits. The ones digit is 2 less than the thousands digit. The hundreds digit is 1 more than the ones digit. The hundreds digit and the tens digit are the same. What is the number on Lilo's card?

Answer ______________________

Strategy ______________________

Write About It

Look back at Problem 6. Describe how the strategy you chose helped you to solve the problem.

____

____

Work Together: Plan a Trip

The members of your group will plan a trip around the world. You will pick three places to visit and figure out how you will get to each place. You will also decide how many days you will stay at each place.

Plan

1. You must begin and end your trip in a city near where you live and make a complete circle around the planet.
2. You must stop to visit at least two continents.
3. You must be away for exactly 80 days.

Decide Your group must agree on three different places to go and how you will travel. You must always travel together as a group.

Create Write a description of your trip. Include the places you will visit in order from beginning to end. Tell how you will travel to each place, how long you will stay, and how you get back.

Present As a group, share your plans with the class. Discuss your reasoning.

UNIT 2 Problem Solving Using Multiplication and Division

Unit Theme:
Having Fun

How do you spend your free time? Some people like to collect things, such as stamps or baseball cards. Other people like to play or listen to music. Many people read books or play sports. In this unit, you will see how math is used in some common activities.

Math to Know

In this unit, you will use these math skills:

- Multiply whole numbers
- Divide whole numbers
- Identify factors and multiples

Problem-Solving Strategies

- Work Backward
- Solve a Simpler Problem
- Make a Table
- Use Logical Reasoning

Link to the Theme

Finish the story. Write about the photo album. Use both words and numbers.

Raj wants to give his mother a photo album for her birthday. He plans to put family photos on each page.

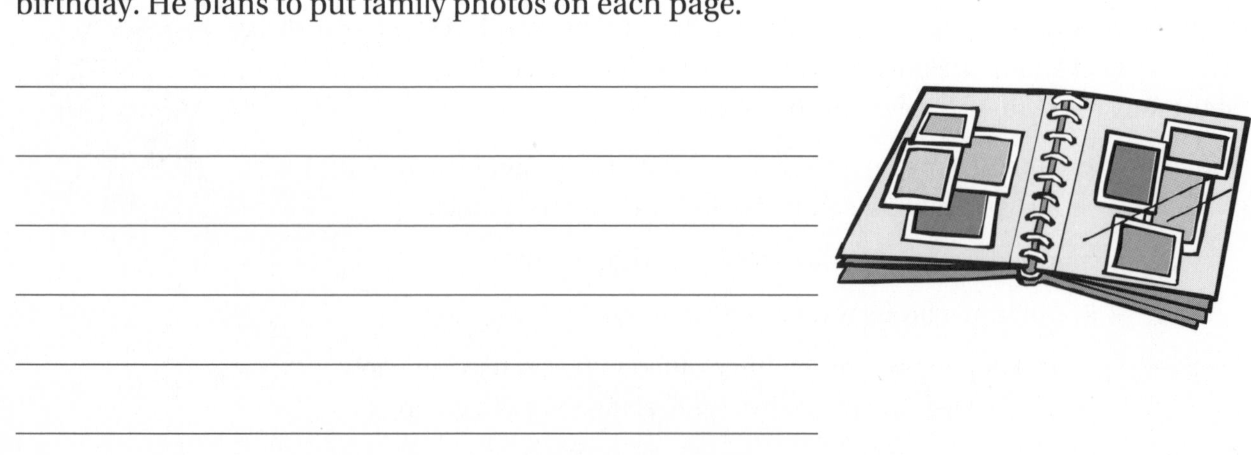

Use Math Language

Review Vocabulary

The list below shows vocabulary terms in this unit. Knowing the meaning of these terms will help you understand the problems.

dividend	factor	product	remainder
divisor	multiple	quotient	Venn diagram

Vocabulary Activity **Multiple-Meaning Words**

Some terms have more than one meaning. Use terms from the list above to complete the following sentences.

1. The ____________ of 3 and 4 is 12.
2. Before buying a new bike, Tai read a review of the ____________ .
3. 9 is a ____________ of 3.
4. There are ____________ , or many, ways to solve the same problem.

Graphic Organizer **Word Map**

Complete the graphic organizer.

- Write your own definition of *quotient.*
- Draw a picture or diagram to show what the term means.
- Write a number sentence that includes an example of a quotient.
- Write a number sentence that does not include an example of a quotient.

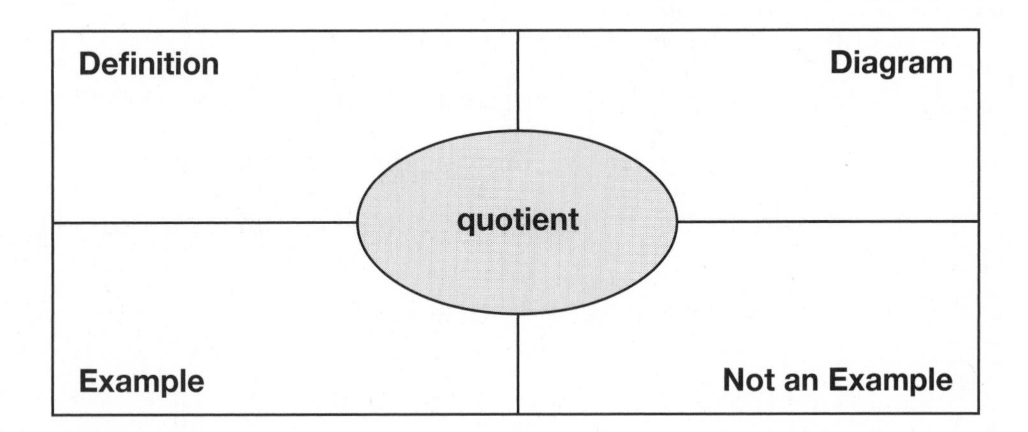

Lesson 5

Strategy Focus Work Backward

MATH FOCUS: Multiplication and Division Facts

Learn About It

Read the Problem

Jin delivers newspapers every day. Each morning, she stacks the newspapers in bundles of 9. She uses 3 feet of string to tie up each bundle. Jin uses 15 feet of string in all. How many newspapers does she deliver each morning?

Reread Ask yourself these questions as you read the problem.

- What is the problem about?

____

- What types of information are given?

____

- What am I asked to find?

____

Mark the Text

Search for Information

Read the problem again. Circle the numbers and important math ideas.

Record What information will help you solve the problem?

Jin stacks the newspapers in bundles of ________ .

She uses ____________ of string for each bundle.

She uses ____________ of string for all the bundles.

You can use this information to choose a problem-solving strategy.

Decide What to Do

You know the number of newspapers in each bundle. You know how much string Jin uses to tie each bundle. You know how much string Jin uses to tie all the bundles.

Ask How can I find the number of newspapers Jin delivers each morning?

- I can use the strategy *Work Backward.*
- I can begin with the total length of string Jin uses to tie all the bundles. Then I can work backward to the start.

Use Your Ideas

Step 1 Make a diagram to show what happens in the problem.

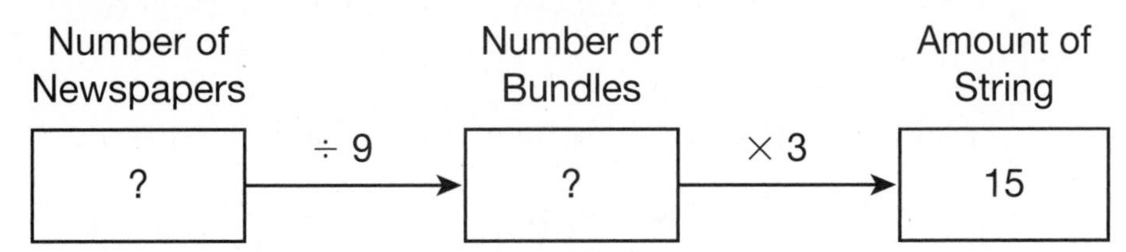

Step 2 Use the diagram to help you work backward.

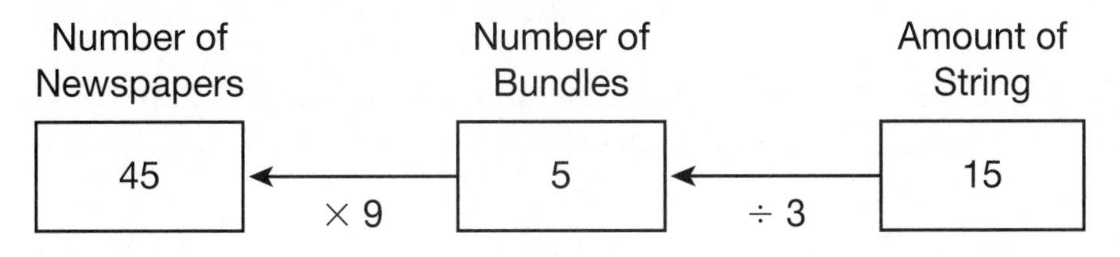

To work backward, undo what has happened in the problem. To undo an operation, use the opposite operation.

First, divide 15 by 3 to find the number of bundles. There are ________ bundles.

Then multiply that answer by ________ to find the total number of newspapers.

So Jin delivers ________ newspapers each morning.

Review Your Work

Check your answer by working from the start.

Describe How does the second diagram show that you are working backward?

____

____

Try It

Solve the problem.

① Students are collecting books for the senior center. Jasper collected twice as many books as Alice. Alice collected 3 more books than Ed. Jasper collected 16 books. How many books did Ed collect?

Read the Problem and Search for Information

Reread the problem. Circle numbers or words you need to solve the problem.

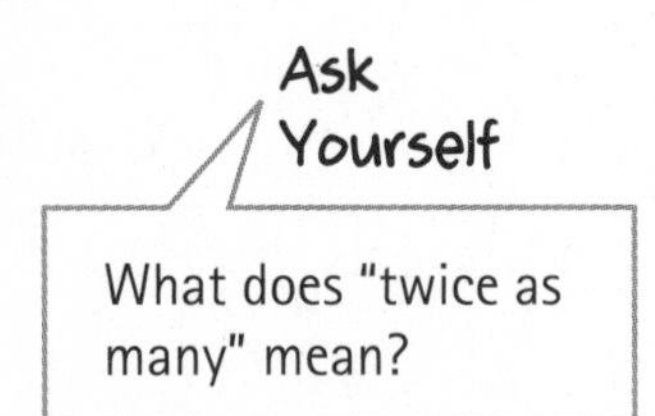

Decide What to Do and Use Your Ideas

You can use the strategy *Work Backward* to find the number of books Ed collected.

Step 1 Make a diagram to show the information in the problem.

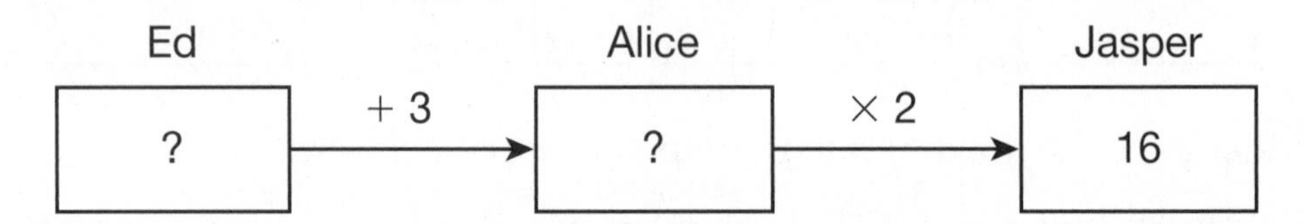

Step 2 Use the diagram to help you work backward. Fill in the number of books at each step.

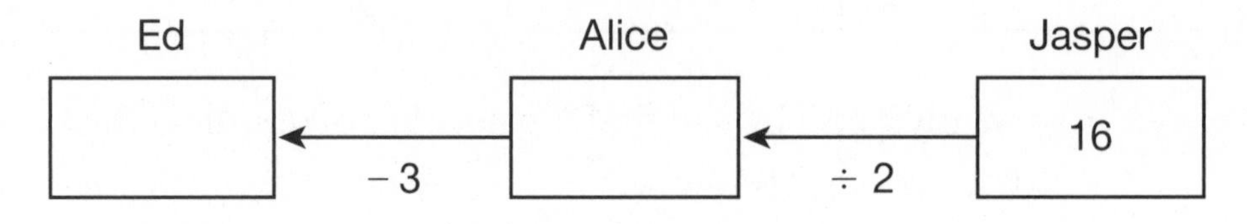

So Ed collected ________ books.

Review Your Work

Go back to the start of the problem. Make sure all the information in the problem fits with your answer.

Explain Why do you divide the number of books Jasper collected by 2 to find the number that Alice collected?

____

____

Apply Your Skills

Solve the problems.

② Della is counting the books in her bookcase. She has 3 times as many science fiction books as mystery books. She has 2 more mystery books than comic books. She has 27 science fiction books. How many comic books does Della have?

> **Ask Yourself**
> What does "3 times as many" mean?

Della has ________ science fiction books.

She has ________ times as many science fiction books as mystery books.

> **Hint** Work backward from the number of science fiction books Della has.

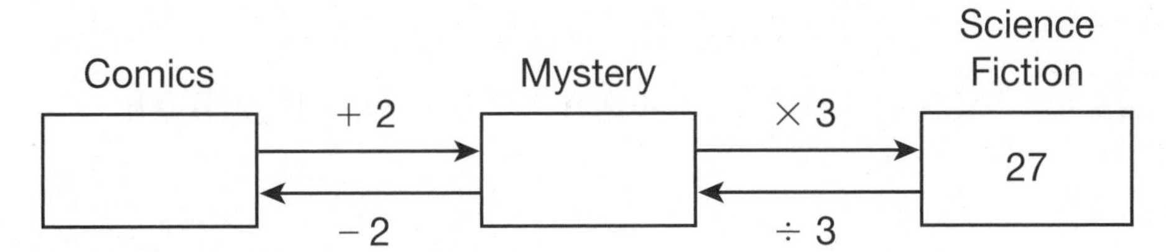

Answer ________________________________

Summarize How did working backward help you solve the problem?

____

____

③ Mrs. Pina is using white paper and pink paper to make a scrapbook. She makes several sections, each with 6 sheets of white paper. She makes another section that has 14 sheets of pink paper. There are 32 sheets of paper in the scrapbook altogether. How many sections have white paper?

> **Ask Yourself**
> How can I use a diagram to work backward?

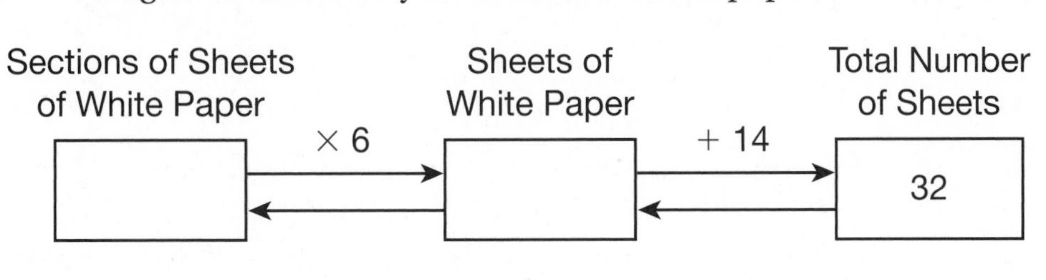

> **Hint** To undo an operation, use the opposite operation. Write the opposite operations on the diagram.

Answer ________________________________

Sequence Explain the steps you would use to check your answer.

____

____

4 There are 10 bookshelves in the children's room of the bookstore. Each shelf has the same number of books. There is also a sale table with 30 books. There are 120 books altogether in the children's room. How many books are on each of the 10 shelves?

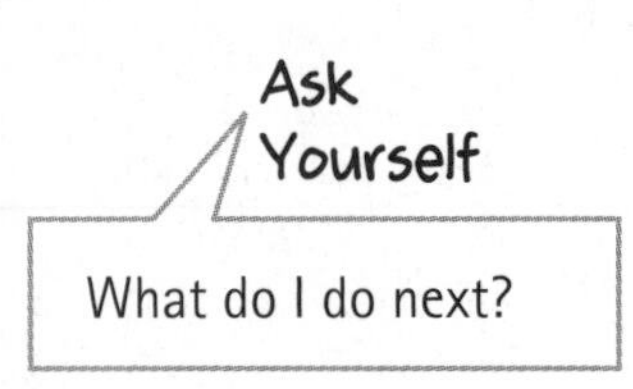

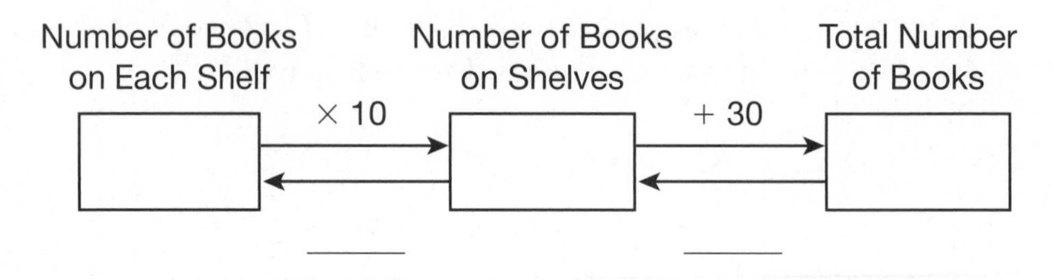

Ask Yourself

What do I do next?

Answer ______________________

Defend Explain why the top arrow on the left of the diagram shows × 10.

____

____

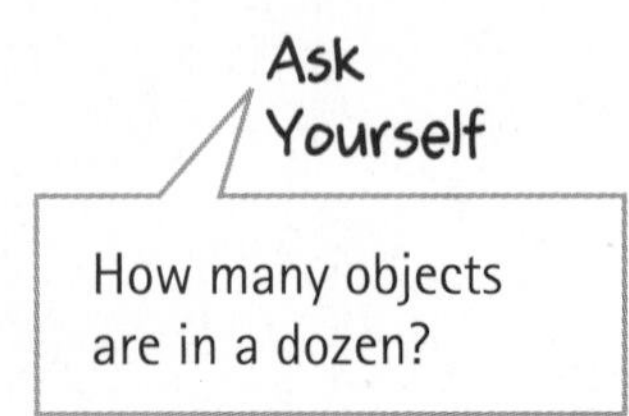

5 Nick packs the same number of comic books in 2 boxes. He gives his friend a dozen comic books from one box. Now Nick has 18 comic books in that box. How many comic books did Nick have in all before he gave some to his friend?

Hint You need to use more than one operation to find the answer to this problem.

Nick has ________ comic books left in one box.

He gave his friend ________ comic books.

Answer ______________________

Conclude Suppose Ty finds the answer to be 30 comic books. What was Ty's error?

____

____

On Your Own

Solve the problems. Show your work.

6. Sydney puts books back on shelves at the library. She left for the day at 3:00 P.M. with 15 books still on the cart. Before Sydney left, she put back 5 books on each of 4 shelves. How many books were on the cart when Sydney started putting books away?

Answer ______________________________

Decide What information is given that is not needed to solve the problem?

____

____

7. Ms. Bell has four times as many note cards as blank cards. She has 4 fewer blank cards than birthday cards. Ms. Bell has 10 birthday cards. How many note cards does Ms. Bell have?

Answer ______________________________

Infer Why might you start at the end of this problem to help you solve it?

____

____

Create

Look back at Problem 6. Change the number of books left on the cart and the number of shelves. Write and solve a problem about a different student using the numbers you have chosen.

Strategy Focus

Solve a Simpler Problem

MATH FOCUS: Multiplication by 1- and 2-Digit Numbers

Learn About It

Read the Problem

Nine students are trying out for a school musical. Students will try out in pairs. Each student will sing once with each of the other eight students. Each pair will sing for three minutes. How long will the tryouts take?

Reread Ask yourself questions about the problem.

- What is the problem about?

____

- What do you notice about all the numbers in this problem?

____

- What are you asked to find?

____

Mark the Text

Search for Information

Underline the number words in the problem. Decide which numbers are important.

Record What information will help you solve this problem?

The students will try out in pairs. This means they will try out ________ at a time.

Each student will sing ____________ with each of the other students.

Each pair will sing for ________ minutes.

Use this information to help you decide on a strategy.

Decide What to Do

You know how long each tryout will take. You know how many times each student will sing.

Ask How can I find how long it will take for all the pairs to sing?

- I can use the strategy *Solve a Simpler Problem.* I can solve the problem using fewer students. Then I can see what to do for more students.

Sometimes a problem seems difficult to solve. If you solve a simpler problem first, you will often see how you can solve the more difficult problem.

Use Your Ideas

Step 1 Instead of 9 students, look at how up to 5 students can be paired. Each dot is a student. Each line shows a pair.

Number of Students	2	3	4	5
Number of Tryouts	1	3	6	10

Step 2 Make a table. Continue the pattern. Find the number of tryouts for 9 students.

Number of Students	2	3	4	5	6	7	8	9
Number of Tryouts	1	3	6	10	15	21	28	

+2 +3 +4 +5 +6 +7 + ____

Step 3 Multiply the number of tryouts by 3 minutes. The product tells you the total amount of time the tryouts will take.

So the tryouts will take ________ minutes.

Review Your Work

Check your answer to make sure it makes sense.

Explain Why is solving a simpler problem a good strategy for this problem?

____

____

Try It

Solve the problem.

① AJ and his friends are donating CDs to a yard sale the Music Club is having. AJ donates 6 CDs. His friends donate 8, 10, 12, 14, 16, 18, 20, 22, and 24 CDs. At the yard sale, all the CDs sell for $4 each. How many CDs were donated in all?

Read the Problem and Search for Information

Find the information you need to solve the problem. Some information may not be needed.

Decide What to Do and Use Your Ideas

You can use the strategy *Solve a Simpler Problem* to think about an easier way to add the 10 numbers.

Step 1 List all the numbers. See if any pairs of numbers have sums that are multiples of 10. Mark those pairs.

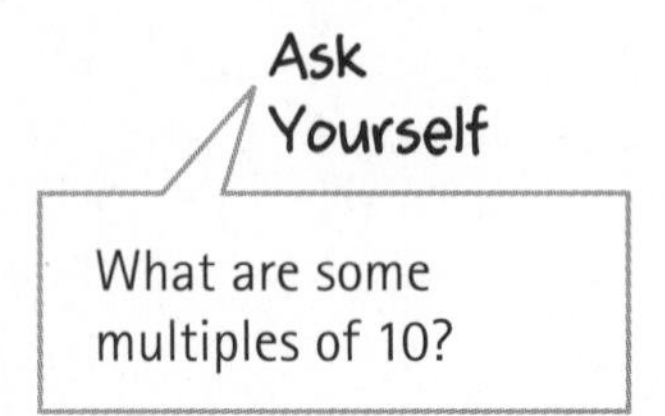

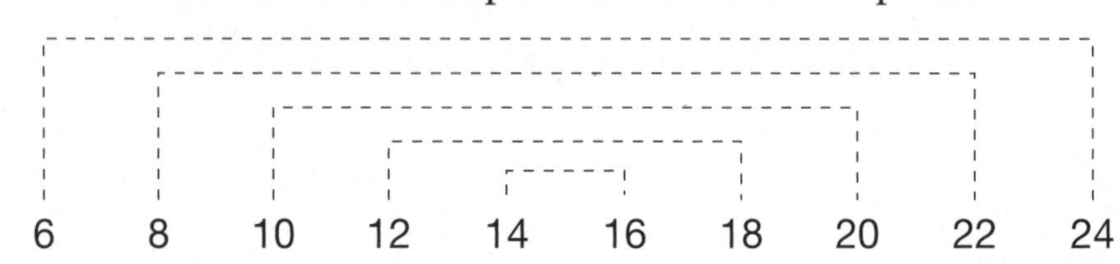

Step 2 Find the sum of each pair.

$6 + 24 =$ ________ $8 + 22 =$ ________ $10 + 20 =$ ________

$12 + 18 =$ ________ $14 + 16 =$ ________

Step 3 Find the product for 5 groups of 30.

$5 \times 30 =$ ________

So ________ CDs were donated in all.

Review Your Work

Make sure you answered the question in the problem.

Describe Why are multiples of 10 simpler numbers to work with when finding a sum or product?

____

____

Apply Your Skills

Solve the problems.

② Terri and Pablo help set up tables for the Jazz Awards. In each row, they will push together 8 tables that are the same size. One table seats 4 people. When pushed together, 2 tables will seat 6 people as shown below. How many chairs will they need if they set up a total of 5 rows?

Drawing of Tables and Chairs	× ×□× ×	× × ×□□× × ×		
Number of Tables	1	2		
Number of Chairs	4	6		

+2   + ____   + ____

Hint Draw a picture to show 3 tables pushed together and 4 tables pushed together.

Ask Yourself

How many chairs are there for 5 tables pushed together? For 6 tables pushed together?

The number of chairs needed for 1 row is ________ .

Answer ____________________

Determine How does making a drawing make this a simpler problem to solve?

____

____

③ Eva has a new MP3 player. The first day she downloads 1 song to her MP3 player. The next day she downloads 2 songs. On the third day she downloads 3 songs. She continues to download 1 more song every day. How many songs are on her MP3 player at the end of 20 days?

Hint Find pairs of numbers that have a sum of 20.

1 2 3 4 5 6 7 8 9 10 11 12 13 14 15 16 17 18 19 20

1 + 19 = ________   2 + 18 = ________   3 + 17 = ________

Will I pair every number?

Answer ____________________

Conclude Suppose Eva downloaded the same number of songs every day for 20 days. How could you solve that problem?

____

____

Hint Mark the numbers that make good pairs.

Ask Yourself

What operations do I need to use to find the number of people marching in the band?

Ask Yourself

Can I use a pattern to solve this problem?

Hint Sketch a 4-step staircase.

④ A marching band is practicing in the field. The people march in different rows. The number of people in each row is listed below. How many people are marching in the band?

1 3 5 7 9 11 13 15 17 19 21

1 + 19 = ________

Answer ______________________

Analyze May paired numbers that have sums of 24. Will her strategy work? Explain.

____

____

⑤ A stage crew uses blocks to build a staircase for the school musical. Each block costs $3. A 1-step staircase uses 1 block. A 2-step staircase uses 3 blocks. A 3-step staircase uses 6 blocks, and so on. How much will it cost to build a 7-step staircase?

1 block   3 blocks   ____ blocks

Answer ______________________

Compare Find another problem in this lesson that is like this problem and explain why the two are similar.

____

____

On Your Own

Solve the problems. Show your work.

6. There are 6 elementary schools. Each school has a chorus. During the year, each chorus performs with every other chorus twice. How many performances are there in a year?

Answer ______________________________

Assess Lee found the number of performances to be 15. What was Lee's error?

____

____

7. A flute player will play 23 concerts this year. She plans to play 2 fewer concerts each year until the year she plays only 1 concert. Then, she will retire. She will save $600 of the money she makes from each concert in a special account. How much money will she put in that account?

Answer ______________________________

Revise Describe the method you used to solve the problem and explain how you could have solved it in a different way.

____

____

Create

Look back at Problem 2. Write and solve a problem about tables that seat a different number of people and with a different number of tables pushed together.

Strategy Focus

Make a Table

MATH FOCUS: Division by 1-Digit Numbers

Learn About It

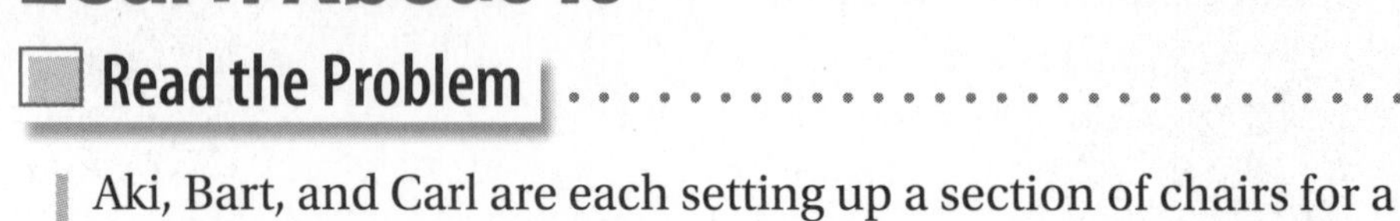

Read the Problem

Aki, Bart, and Carl are each setting up a section of chairs for a talent show. At least 250 people will attend. Aki has 114 chairs to put in rows of 6. Bart has 109 chairs to put in rows of 9. Carl has 74 chairs to put in rows of 8. How many rows can each person make? How many chairs will each person have left over? Are enough chairs left over to make another row in any section?

Reread Use your own words to describe the problem.

- What is this problem about?

____

- What information does the problem give?

____

- How many questions does the problem ask? What are they?

____

Search for Information

Reread the problem. Look for information to help you solve it.

Record Find details you will need to solve this problem.

Aki has ________ chairs to set up in rows of ________ .

Bart has ________ chairs to set up in rows of ________ .

Carl has ________ chairs to set up in rows of ________ .

Use these details to help you decide on a strategy.

Decide What to Do

There are two kinds of information you need to use: the *number of chairs each person has* and the *number of chairs in each row.*

A table can help you organize the information.

Ask How can I find the number of rows each person can make, the number of leftover chairs, and whether another row can be made?

- I can use the strategy *Make a Table* to organize the data.
- I can divide. The dividend is the number of chairs the person has. The divisor is the number of chairs in each row. The quotient tells the number of rows. The remainder tells how many leftover chairs the person has.
- I can add the number of chairs left over from each person. Then I can see if there are enough to make another row in any section.

Use Your Ideas

Step 1 Divide to find the number of rows in each section and the number of leftover chairs. Record the results in the table.

	Number of Chairs	Chairs in Each Row	Number of Rows	Chairs Left Over	
Aki	114	6	19	0	← 114 ÷ 6 = 19 R0
Bart	109	9	12		← 109 ÷ 9 = 12 R ________
Carl	74	8			← 74 ÷ 8 = ________ R ________
			Sum →		

Step 2 Add the numbers of chairs left over to see if there are enough to make one more row.

Altogether, there are ________ chairs left over. There ____________ enough chairs to make another row.

Review Your Work

Be sure you have answered all the questions that were asked.

Relate Which words in the problem tell you there may be remainders when you divide?

____

____

Try It

Solve the problem.

① The 4th graders and 5th graders are going on a field trip. The eighty-five 4th graders are arranged in groups of 5. The ninety-six 5th graders are arranged in groups of 8. The 4th-grade groups need 1 adult for each group. The 5th-grade groups are larger, so they need 2 adults for each group. How many adults are needed for the trip?

Read the Problem and Search for Information

Identify what you need to find. Circle numbers and words that will help you solve the problem.

Decide What to Do and Use Your Ideas

You can use the strategy *Make a Table* to help you organize information. Write what you know in this table.

Ask Yourself

If I know the number of students and the number of students in a group, how can I find the number of groups?

Grade	Number of Students	Students in 1 Group	Number of Groups	Adults in 1 Group	Adults Needed
4th	85	5		1	
5th	96	8		2	
				Sum →	

Step 1 Divide to find the number of groups in each grade.

Step 2 Multiply to find the number of adults needed.

Step 3 Find the total number of adults needed for the trip.

So ________ adults are needed for the trip.

Review Your Work

Check that you have answered the question that was asked.

Explain How does making a table help you solve this problem?

____

____

Apply Your Skills

Solve the problems.

Ask Yourself

What operations can I use to solve the problem?

② Mr. Wu orders snacks to sell at the drama club's Spring Fling. He orders 152 granola bars. The granola bars are packed in cartons that hold 8 bars each. He also orders 324 boxes of raisins. The raisins are packed in cartons that hold 6 boxes each. How many cartons of snacks does Mr. Wu order?

Snacks	Number of Snacks	Snacks in 1 Carton	Number of Cartons
Granola Bars			
Boxes of Raisins			
		Sum →	

Hint You need to find how many cartons it takes to pack all the snacks.

Answer ____________________

Identify What operations did you use to fill in the table?

____

____

③ A dance supply store ships 120 pairs each of three types of shoes: ballet, ballroom, and tap. Ballet shoes are packed 8 pairs to a case. Ballroom shoes are packed 4 pairs to a case. Tap shoes are packed 6 pairs to a case. It costs $7 to ship one case. How much will it cost to ship all of the cases?

Type of Shoe	Number of Pairs	Pairs in 1 Case	Number of Cases	Cost to Ship 1 Case	
Ballet					
Ballroom					
Tap					

Hint Use the question to help you decide how to label the last column of the table.

Answer ____________________

Ask Yourself

How can I find the cost to ship each type of shoe?

Demonstrate Write about how you could solve this problem without using the last two columns of the table.

____

____

Hint Underline the words in the problem that will help you label the first column of the table.

Ask Yourself

Which column of numbers will I compare to help me answer the question?

④ At a school concert, Ashley sells adult tickets for $9 each. She collects $1,449. Patrick sells student tickets for $4 each. He collects $392. Eduardo sells senior citizen tickets for $6 each. He collects $1,020. Who sold the most tickets?

Type of Ticket	Amount Collected	Price for 1 Ticket	Number of Tickets Sold

Answer ______________________________

Interpret Why did the person who sold the most tickets not collect the most money?

____

____

Ask Yourself

What do I need to know to be able to answer the question?

Hint One column of your table will have the same number in every space.

⑤ To raise money, 3 members of a dance club sell T-shirts. They want to sell a total of 150 T-shirts for $9 each. So far, by selling the T-shirts, Sandra has raised $279, Jae has raised $369, and Marta has raised $171. How many more T-shirts do they need to sell?

Answer ______________________________

Decide What other question could you ask about the T-shirts sold or the money raised by the dance club?

____

____

On Your Own

Solve the problems. Show your work.

6. Fliers for a school play will be displayed around town. Kelly has agreed to post 78 fliers. Kyle will post 42 fliers. Paulo will post 12 fliers. Manny will post 36 fliers. The school buys fliers in packages of 6. How many packages does the school need to buy?

Answer ______________________________

Consider Describe another way of finding how many packages the school needs to buy.

____

____

7. Last weekend the new movie *Otters in Outer Space* opened in theaters. City Cinema charged $9 for a ticket and made $11,142. The Corner Cinema charged $8 for a ticket and made $11,168. Which cinema sold more tickets? How many more?

Answer ______________________________

Determine You needed to use two operations to solve this problem. Which phrases helped you decide which operations to use?

____

____

Create

Look back at Problem 4. Change the ticket prices and the amount of money each student collects. Write and solve your problem.

Strategy Focus

Use Logical Reasoning

MATH FOCUS: Multiples, Factors, Prime, and Composite Numbers

Learn About It

Read the Problem

Jo has a collection of mini-erasers. Ann asks her how many she has. Jo says that the number is the greatest number that matches all these clues:

It is not more than 16. One of the factors of the number is 2. The number is a multiple of 3.

How many mini-erasers does Jo have?

Reread Ask yourself questions as you read.

- What is the problem about?

____

- What kind of information is given?

____

- What are you asked to do?

____

Mark the Text

Search for Information

Reread the problem and look for words and numbers that will help you find a solution.

Record What information do you know about the number of erasers?

The number is not more than ________ .

One factor of the number is ________ .

The number is a ______________ of 3.

You can use this information to solve the problem.

Decide What to Do

The problem gives you clues to find the number of erasers. You can use the clues one at a time to find the answer.

Ask How can I find the number of mini-erasers in Jo's collection?

- I can use the strategy *Use Logical Reasoning* to solve this problem.
- I can find the number that matches all the clues. A Venn diagram can help me organize the numbers.

Use Your Ideas

Step 1 List numbers up to 16 that have 2 as a factor.

2, 4, 6, 8, 10, 12, ________, ________

Step 2 List numbers up to 16 that are multiples of 3.

3, 6, 9, ________, ________

Step 3 Organize the numbers from Step 1 and Step 2.

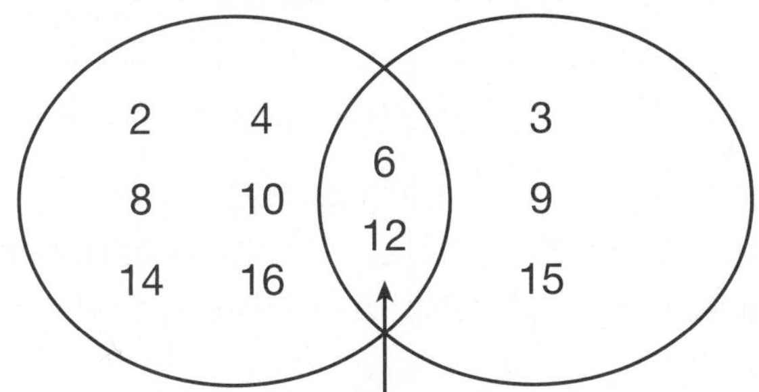

Which numbers have 2 as a factor *and* are multiples of 3? ____________

So Jo has ________ mini-erasers.

The number of mini-erasers is the *greatest* number that matches all the clues.

Review Your Work

Reread the problem to see if you answered the question asked.

Clarify How did you use logical reasoning to solve the problem?

____

____

Try It

Solve the problem.

① Yin, Tom, Sarah, and Jason each have a marble collection. In no particular order, they have 7, 16, 20, and 23 marbles. The number of marbles Yin has is a multiple of 2. Sarah has 3 fewer marbles than Tom. How many marbles does Jason have?

Read the Problem and Search for Information

Identify what is given and what you are asked to find. Reread the problem and circle the words and numbers you need.

Decide What to Do and Use Your Ideas

You can use the strategy *Use Logical Reasoning* to solve this problem. Make a chart to rule out possible answers.

	7	16	20	23
Yin				
Tom				
Sarah				
Jason				

Ask Yourself

Which clue should I use first?

Step 1 Use the clue: *Sarah has 3 fewer marbles than Tom.*

Sarah has ________ marbles and Tom has ________ because $23 - 20 = 3$. Use ✔s and Xs in the empty boxes in those rows and columns to rule out other choices.

Step 2 The numbers that are left to choose from are ________ and ________. Only one of those numbers is a multiple of 2. Yin has ________ marbles. Use ✔s and Xs to show this.

So Jason has ________ marbles.

Review Your Work

Make sure you answered the question that was asked.

Describe How do you know which clue to use first?

____

____

Apply Your Skills

Solve the problems.

② On opening day at a movie theater, every fourth person gets a free set of collectible cards. Every seventh person gets a free movie ticket. There are 40 people in line. How many people will get both the cards and the ticket?

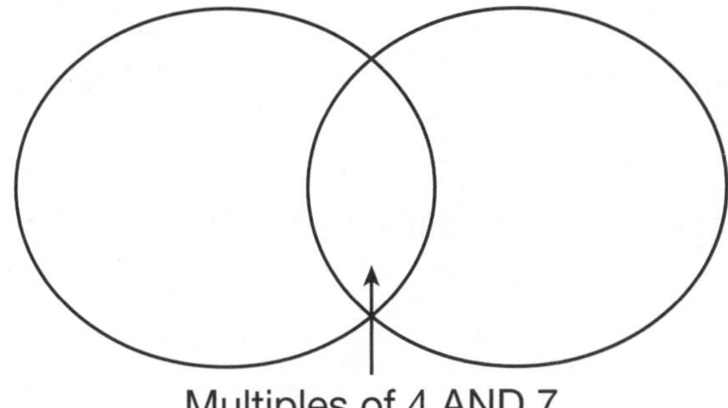

Answer ______________________________

Explain How does using a Venn diagram help you to solve the problem?

____

____

Ask Yourself

Which numbers are multiples of 4? Which numbers are multiples of 7?

Hint Use the Venn diagram. The numbers you write in the overlapping part show the answer to the problem.

③ Juan, Gina, Cory, and Hans had a contest to see who could read the most books over the summer. They read 3, 8, 9, and 13 books. The number of books Cory read is a prime number. The number of books Gina read is three times the number that Hans read. Who won the contest?

	3	8	9	13
Juan				
Gina				
Cory				
Hans				

Answer ______________________________

Determine How did you know how many books Cory read?

____

____

Ask Yourself

Which number is three times another number?

Hint Put a ✔ in a box to show the number of books for each friend. Put an X in the rest of the boxes in the row.

Ask Yourself

Which two numbers have a common multiple of 24?

Hint Find the number of dolls in Gail's bedroom to help you find the number of dolls in the guest bedroom.

④ Gail collects dolls. She has doll groups of 5, 8, 10, and 12 in each of 4 rooms. The number of dolls in Gail's bedroom and the number of dolls in the living room have a common multiple of 24. There are 3 more dolls in Gail's bedroom than in the dining room. How many are in the guest bedroom?

Dining Room				
Gail's Bedroom				
Living Room				
Guest Bedroom				

Answer ______________________________

Sequence How did you use the two clues to find the number of dolls in each room?

____

____

Hint The answer to the problem is one number.

Ask Yourself

How can I find multiples of 5 and 6?

⑤ There are 60 customers at the grand opening of Choo-Choo Trains. Every fifth customer gets a conductor's hat. Every sixth customer gets a model caboose. Which number customer is the first to get both gifts?

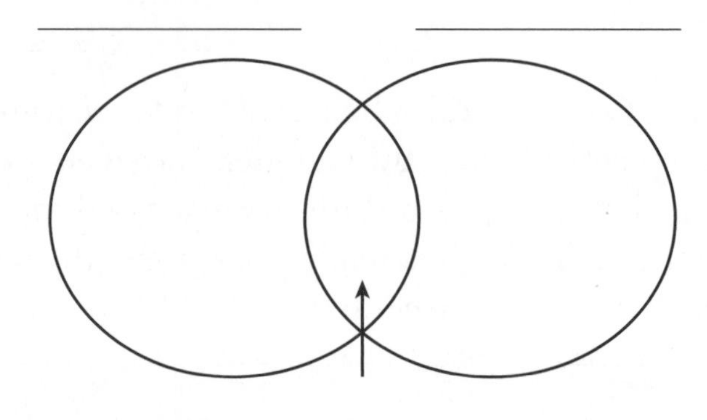

Answer ______________________________

Consider Do you need to write all of the multiples for 5 and 6 up to 60? Explain.

____

____

On Your Own

Solve the problems. Show your work.

(6) Jan collects toy animals. She has bears, monkeys, elephants, and frogs. The number she has of each type is 6, 7, 14, and 21. The number of frogs is a prime number. The number of monkeys is a multiple of 3. The number of bears is three times the number of frogs. How many elephants does Jan have?

Answer ______________________________

Infer Why do you not need a clue about elephants to find how many Jan has?

____

____

(7) The Toy Chest is giving a prize to the customer who knows the correct number of action figures on a hidden display. The Toy Chest gives these clues:

The number is less than 30. One of the factors of the number is 3. The number is a multiple of 4. The number is greater than 12.

How many action figures are on the hidden display?

Answer ______________________________

Assess Molly says that the clue that the number is greater than 12 is not needed. Do you agree with Molly? Explain why or why not.

____

____

Create

Look back at Problem 2. Change the numbers of the people who receive the free cards and tickets. Write and solve a problem about the number of people who get gifts.

UNIT 2 Review What You Learned

In this unit, you worked with four problem-solving strategies. You can often use more than one strategy to solve a problem. So if a strategy does not seem to be working, try a different one.

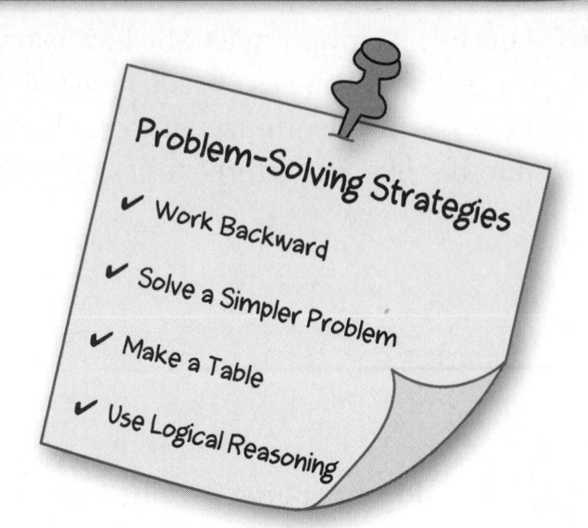

Solve each problem. Show your work. Record the strategy you use.

1. Cara is saving to buy a book that costs $23. She is paid $3 each time she walks the neighbor's dog. If Cara walks the dog 4 times, she will have exactly enough money to buy the book. How much does she already have saved?

Answer ______________________________

Strategy ______________________________

2. The Listen Up Music store sells different kinds of CDs. Used CDs are $4 each. New releases are $14 each. Oldies are $8 each. On Thursday, the store sold 40 used CDs, 25 new releases, and 17 oldies. What were the store's total sales that day?

Answer ______________________________

Strategy ______________________________

3. There are 10 dancers and 8 drummers waiting to perform. Each dancer shakes hands with every other dancer. Each drummer shakes hands with every other drummer. How many more handshakes are there between dancers than between drummers?

Answer ______________________________

Strategy ______________________________

4. Ana raises worms for gardens. She bought some worms in February. She had twice as many worms by April. She had 750 worms in June. That is 3 times as many worms as she had in April. How many worms did Ana buy in February?

Answer ______________________________

Strategy ______________________________

5. Phil, Matt, Julie, and Kat are on the same soccer team. Their jersey numbers are 14, 16, 23, and 27. Matt's number is a multiple of 4. Phil's number is 9 less than Julie's number. What is Kat's number?

Answer ______________________________

Strategy ______________________________

Explain how you used the clues to solve the problem.

Solve each problem. Show your work. Record the strategy you use.

6. Mason writes a secret number on a piece of paper and folds it. He gives these clues about his number:

The number has both 3 and 5 as factors. It is less than 40. The number is not 15.

What is Mason's number?

Answer ______________________________

Strategy ______________________________

7. A tour group ordered 14 adult tickets for $35 each and 8 student tickets for $20 each. The group also paid a $2 fee for each ticket ordered. What was the total cost of the tickets?

Answer ______________________________

Strategy ______________________________

8. A nursery school buys 32 round blocks and 16 square blocks. These blocks come in large boxes. Each box holds 16 blocks and costs $45. The school also buys 24 rectangular blocks and 48 triangular blocks. These blocks come in small boxes. Each of these boxes holds 8 blocks and costs $30. How much does the school pay in all for the blocks?

Answer ______________________________

Strategy ______________________________

You may have made a table with 4 rows to help you solve this problem. Explain how you could solve this problem using a table that has only 2 rows.

9. Marco, Eva, and Pedro collect baseball cards. Marco has 12 more cards than Eva. Eva has 8 more cards than Pedro. Pedro has 16 cards. How many baseball cards does Marco have?

Answer ______________________________

Strategy ______________________________

10. An artist sells hand-painted cards. He sells 4 bird cards at $1 each, 12 cat cards at $2 each, 16 city cards at $1 each, 11 dog cards at $2 each, and 18 fish cards at $1 each. How much money does the artist make?

Answer ______________________________

Strategy ______________________________

Write About It

Look back at Problem 4 and explain the reasoning you used to solve the problem.

____

____

Work Together: Plan a Budget

Your class votes to collect 500 books for a homeless shelter. The class has $200 to spend. A local bookstore will sell books to your class at a low price. Your class can ask for books from students in your school and from local companies. The class plans to print both color and black-and-white fliers for publicity. How will you use the $200?

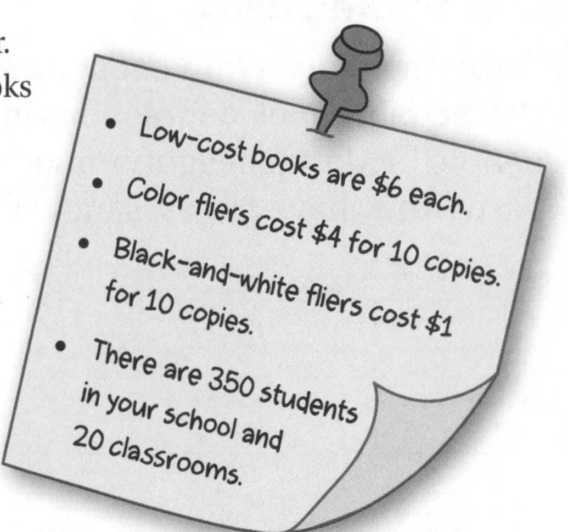

Plan

1. As a group, discuss your ideas about the best way to get the most books. How might the fliers help?
2. Think about what kind of fliers to use. Decide whether you want more than one kind of flier.
3. Decide who should receive the fliers. Find how many copies of each you will need.
4. Decide how much money you want to spend on low-cost books.

Create Outline two different budget-spending plans.

Analyze Choose the plan that will best help you reach your goal.

Present Share your plan with the class. Explain how you will use the $200.

UNIT 3 Problem Solving Using Fractions and Decimals

Unit Theme:
Neighborhoods

What does your neighborhood look like? There is probably a lot to see. You may see kids playing sports and people walking their dogs. No matter where you live, you are part of a neighborhood. In this unit, you will see how math is used right outside your door.

Math to Know

In this unit, you will use these math skills:

- Apply number sense about fractions and decimals
- Add and subtract fractions
- Add and subtract decimals

Problem-Solving Strategies

- Draw a Diagram
- Work Backward
- Use Logical Reasoning
- Look for a Pattern

Link to the Theme

Finish the story. Write about how the food will be shared fairly. Include fractions in your story.
People in Maria's neighborhood are holding a block party. They have pizzas, sandwiches, cakes, and pies.

____

____

____

____

____

____

____

Use Math Language

Review Vocabulary

The list below shows vocabulary terms in this unit. Knowing the meaning of these terms will help you understand the problems.

decimal	denominator	fraction	numerator
decimal point	equivalent	hundredths	tenths

Vocabulary Activity **Word Roots**

Word roots can provide clues to the meaning of some math terms. Use two words from the list above to complete the following sentences.

1. The word ________________ has the same word root as *fracture.* If you fracture something, you break it or separate it into parts.
2. A ________________ names part of a whole or part of a group.
3. An equilateral triangle has sides that are all the same length. The word ________________ has the same word root as *equilateral.*
4. If two fractions have the same value, they are ________________ fractions.

Graphic Organizer **Word Web**

Complete the graphic organizer.

- In the top circle, give an example of a decimal.
- In each of the 3 smaller circles, write a vocabulary word that is related to the word *decimal.*

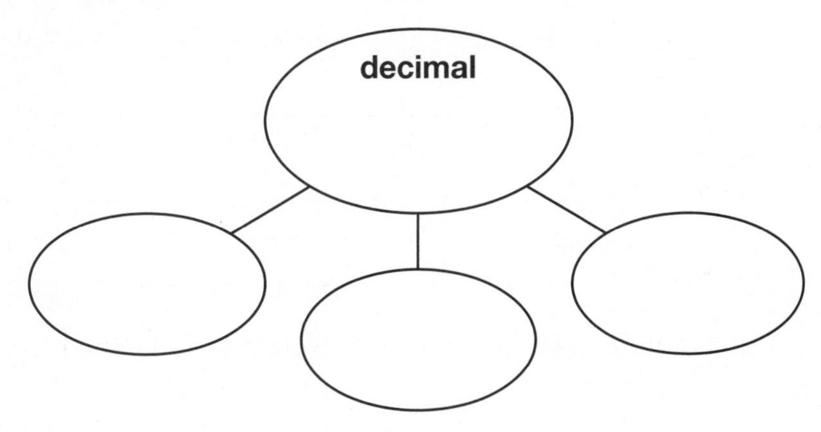

Strategy Focus

Draw a Diagram

MATH FOCUS: Fraction Concepts

Learn About It

Read the Problem

The soccer league has 20 teams. Each team has played the same number of games. Players from three teams are comparing their team records. The Jets won $\frac{2}{3}$ of their games. The Stars won 7 out of 12 games. The Bells won 5 out of every 6 games they played. Which of these teams has the best record?

Reread Ask yourself questions as you read the problem again.

- What is happening in the problem?

____

- In what ways did the players describe their team records?

____

- What question am I asked to answer?

____

Mark the Text

Search for Information

Read the problem again. Circle the information about each team.

Record Write what you know about each team record.

Jets won ______________________________.

Stars won ______________________________.

Bells won ______________________________.

Think about how you can use this data to solve the problem.

Decide What to Do

The data show the team records. You want to compare the team records, but the data are in different forms.

Ask How can I find which of the teams has the best record?

- I can use the strategy *Draw a Diagram* to show each team record.
- I can shade each diagram to show the number of games each team won.
- Then I can compare the shaded parts to find the team with the best record.

Be sure that each diagram is the same size so it will be fair to compare them.

Use Your Ideas

Step 1 Draw a diagram for the Jets with 3 equal parts. Shade 2 parts to show $\frac{2}{3}$ of games won.

Step 2 Draw a diagram for the Stars with 12 equal parts. Shade 7 parts to show 7 games won out of 12.

Step 3 Draw a diagram for the Bells with 6 equal parts. Shade ________ parts to show 5 games won out of every 6.

Step 4 Compare the shaded parts of the diagrams. The diagram for the Bells is shaded more than the others.

So the ________ have the best record.

Review Your Work

Check that the correct number of parts is shaded in each diagram.

Determine Why do you make the diagrams the same size, but use different numbers of parts?

____

____

Try It

Solve the problem.

1. Eight basketball teams play in the first round of a contest. The winners of the first round play each other in the second round. The winners of the second round play each other in the final round. The winning team in the final round wins the contest. What fraction of all the games is played in the second round?

Read the Problem and Search for Information

Think about how to find the number of games in each round.

Decide What to Do and Use Your Ideas

You can use the strategy *Draw a Diagram* to show all the games.

Step 1 Call the teams A, B, C, D, E, F, G, and H. Complete the diagram to show the games played in each round.

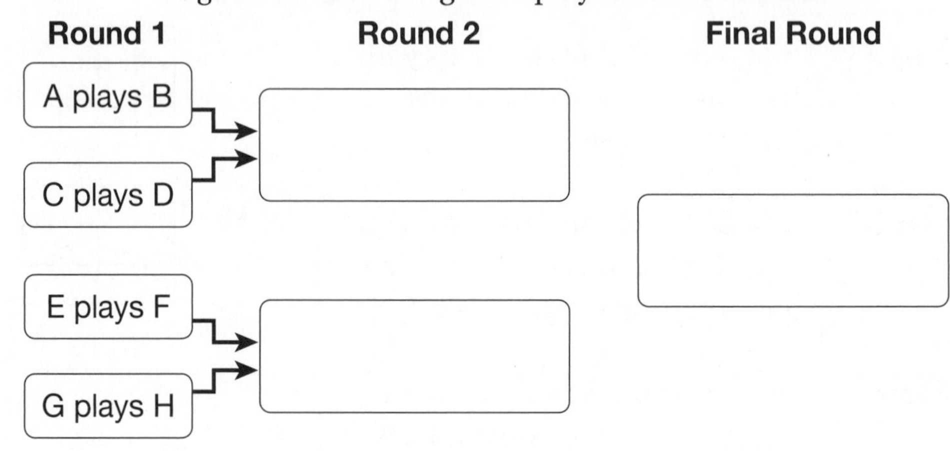

Ask Yourself

Once I know the number of games played in the second round, how can I show that number as a fraction of the total number of games in the contest?

Step 2 Count the games in the second round. ________

Count the total games in the contest. ________

So ________ of the games are played in the second round.

Review Your Work

The number of games should be the numerator of your fraction. Did you use the total of all the games as the denominator?

Explain How did the diagram help you solve this problem?

____

____

Apply Your Skills

Solve the problems.

② The members of a reading club will read the same number of books. Mary has read $\frac{1}{3}$ of the books and Bart has read $\frac{1}{6}$. Rosa has read $\frac{1}{2}$ of the books and Kevin has read $\frac{2}{3}$. What is the order of the readers from most to fewest books already read?

Hint Think about how you should shade and label the diagrams you make.

Ask Yourself What phrase tells me how to order the readers in my answer?

Answer ______________________

Analyze Would your answer change if the readers agree to read 12 or 24 books? Explain.

____

____

③ Ava and Jamal belong to a walking group. They keep journals to show how far they walked. Who walked the farthest distance in one day? On which day did that happen?

Ava	
Mon:	$2\frac{1}{4}$ miles
Wed:	$3\frac{3}{4}$ miles
Fri:	$1\frac{3}{8}$ miles

Jamal	
Mon:	$3\frac{5}{8}$ miles
Wed:	$2\frac{1}{2}$ miles
Fri:	$2\frac{7}{8}$ miles

Ask Yourself What numbers do I need to compare?

1 2 3 4

Hint Label the numbers on the number line. Locate the ones you need to compare.

Answer ______________________

Identify Who walked the greater total distance? Explain.

____

____

4. Jack and Abe are playing checkers. They each start with 12 checkers on the board. Abe loses 5 checkers. Jack loses 3 checkers. What fraction of the squares on the board have checkers on them now? Give your answer in simplest form.

Now there are ________ checkers on the board.

Hint Find the number of checkers left on the board now by crossing out the checkers Abe and Jack lose.

Ask Yourself

How can I use the diagram to find the total number of squares?

Answer ______________________________

Decide Sandra said that $\frac{1}{16}$ of the squares have checkers on them. What error did Sandra make?

____

____

5. A new school band was formed in May. There were 5 clarinet players and 2 trumpet players. One person played a trombone. There were also 2 drummers. In June, 2 more people join the band. One plays clarinet. One plays trumpet. How did the fraction of clarinet players in the band change from May to June? Explain your answer.

Hint The number of clarinet players and the total number of band members changed.

Ask Yourself

How many band members were there in May? In June?

Answer ______________________________

Conclude Why do you think the information about trumpet, trombone, and drum players is given in the problem?

____

____

On Your Own

Solve the problems. Show your work.

6. Lee baked fruit bars for her team. She used two pans of the same size. She cut the pan of cherry bars into 12 equal pieces. She cut the pan of apple bars into 8 equal pieces. Lee's team ate 6 pieces of the cherry bars and 5 pieces of the apple bars. Did they eat more of the pan of cherry bars or apple bars? Explain your answer.

Answer ______________________________

Judge Could you answer the question if the bars were not cut into equal pieces?

____

____

7. Many kinds of races will be held at a block party. Josie says she will enter $\frac{3}{4}$ of all the races. Katya will take part in $\frac{5}{6}$ of them. Sean will enter 2 out of every 3 races. Roger says he will be in $\frac{5}{10}$ of the races. Who will enter the least number of races?

Answer ______________________________

Formulate What is another question you can ask about the races?

____

____

Create

Look back at the problems and choose one. Change two of the numbers in the problem to create a new problem. Write and solve this new problem.

Lesson 10

Strategy Focus

Work Backward

MATH FOCUS: Adding and Subtracting Fractions

Learn About It

Read the Problem

Nick played a beanbag toss game at a street fair. He tossed 3 beanbags and hit 3 different parts of a bull's-eye target. His total score was 2 points. The second beanbag landed in the outside circle. The third one landed in the inner circle. How many points did Nick make with the first beanbag?

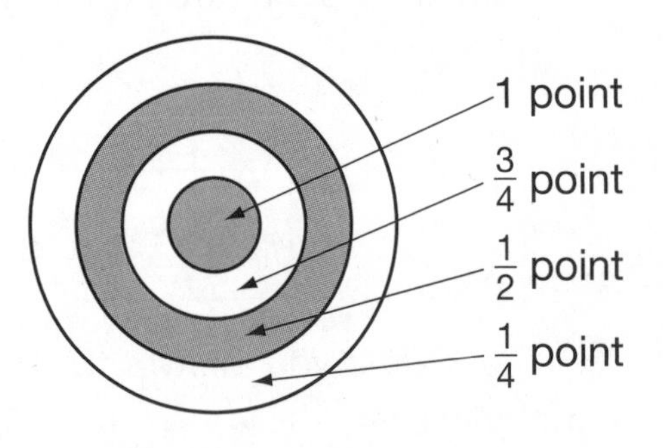

Reread As you read the problem again, study the target drawing.

- What was Nick's total score in the beanbag toss game?

____

- What kinds of information do you see in the problem?

____

- What is the question in this problem?

____

Search for Information

Find information in the problem and in the target drawing.

Record What information can you use to find the answer?

Nick scored ________ points in all.

He scored ________ point with his second toss.

He scored ________ point with his third toss.

You need a strategy you can use to find how many points Nick made with his first toss.

Decide What to Do

The information you recorded can help you find a strategy. You know the total score and the points for 2 tosses.

Ask If the total score is the end result of 3 tosses, how can I trace back to find how many points Nick made with his first toss?

- I can start with the total score. Then I can use the strategy *Work Backward* to find what I want to know.
- I can subtract the points for the second and third tosses from Nick's score. That will tell me what I want to know.

To solve a problem, sometimes you have to start from the last step and trace back to the beginning.

Use Your Ideas

Step 1 Work backward. Start with Nick's total score.

Subtract the points he made with the third beanbag.

$2 - 1 =$ ________

Step 2 Subtract the points he made with the second beanbag from your answer in Step 1.

________ $- \frac{1}{4} =$ ________

Nick scored ________ point with the first beanbag.

Review Your Work

Check your answer by adding the points made with all 3 beanbags to see if the sum is 2 points.

Explain Why is working backward a good strategy for solving this problem?

____

____

Try It

Solve the problem.

1. At her game booth, The Famous Fracto told Roy to think of a fraction. Then she said, "Roy, add 2 to your fraction, subtract $\frac{1}{4}$ from the sum, and tell me your answer." Roy's answer was $2\frac{1}{2}$. The Famous Fracto said, "Roy's fraction is $\frac{3}{4}$." Roy said, "You are right!" How does this number game work?

Read the Problem and Search for Information

Reread to find information that helps you understand how the number game works.

Ask Yourself

How can I undo subtraction? How can I undo addition?

Decide What to Do and Use Your Ideas

Roy used two operations on his fraction. You can use the strategy *Work Backward* to undo those operations.

Step 1 Start with Roy's result, $2\frac{1}{2}$.

First, add $\frac{1}{4}$ to undo subtracting $\frac{1}{4}$.

$2\frac{1}{2} + \frac{1}{4} =$ ________

Step 2 Then subtract 2 to undo adding 2.

This gives you Roy's original fraction.

$2\frac{3}{4} - 2 =$ ________

So the game works by undoing operations, or working backward.

The Famous Fracto found Roy's fraction by adding ________ and subtracting ________ .

Review Your Work

Test your answer by playing the game with another fraction.

Apply Suppose The Famous Fracto asks someone to *multiply* a fraction by 6 and then *divide* that product by 2. How would you undo the operations in this game?

____

____

Apply Your Skills

Solve the problems.

② Cheddar cheese is popular at a cheese store. Kemal cut off and sold chunks from a block of cheese. The chunks weighed $1\frac{1}{2}$ pounds and $2\frac{1}{4}$ pounds. Then he weighed the block of cheese. It weighed $6\frac{3}{4}$ pounds. How much did the block of cheese weigh before Kemal sold those 2 chunks?

Ask Yourself

Will my answer be greater than or less than $6\frac{3}{4}$ pounds?

The weight of the cheese now is ________ pounds.

________ + ________ + ________ = ________

Hint Remember that addition and subtraction undo each other.

Answer ____

Model A diagram would help you see how to work backward. Explain how you would draw a diagram of the problem.

____

____

③ Lee is buying a rectangular frame for a painting at a street fair. To find the cost of the frame, the clerk measures the distance around the painting in inches. Then she multiplies the distance by $1. The charge for the frame is $30.50. The painting is $9\frac{1}{4}$ inches long. How wide is the painting?

Hint First figure out the distance around the painting.

Ask Yourself

What do I know about the lengths of opposite sides in a rectangle?

You can think of $30.50 as $30\frac{1}{2}$ dollars.

The distance around the painting is ________ inches.

The length of the painting is ________ inches.

Answer ____

Conclude How can you use estimation to check if your answer is reasonable?

____

____

Ask Yourself

For this problem, what operations should I use to work backward?

④ A potter sells his flowerpots at a state fair. Someone spilled paint on his sign. The yellow flowerpot is 2 inches taller than the green flowerpot. The blue flowerpot is $2\frac{1}{8}$ inches taller than the yellow one. What is the height of the green flowerpot?

Color	Green	Yellow	Blue
Height	[illegible]	[illegible] in.	$8\frac{1}{4}$ in.
Cost	$11.33	$13.66	$16.00

Hint Find the difference between the heights of the blue and yellow flowerpots first.

$8\frac{1}{4}$ inches − $2\frac{1}{8}$ inches = ____________

Answer ______________________________

Identify What is another question you could ask using the information in this problem?

____

____

⑤ A booth at a street fair sells juice in three different sizes. A small cup holds $5\frac{1}{2}$ ounces, a medium cup holds $7\frac{3}{4}$ ounces, and a large cup holds 10 ounces. Kim bought 1 cup for herself. Then she bought 1 small, 1 medium, and 1 large cup for three of her friends. In all, Kim bought $28\frac{3}{4}$ ounces of juice. What size cup of juice did Kim buy for herself?

Hint Start with the total number of ounces of juice and work backward.

Kim bought ________ cups of juice.

Answer ______________________________

Ask Yourself

Will my answer be a number?

Determine How could you solve this problem using addition and subtraction?

____

____

On Your Own

Solve the problems. Show your work.

6. Lynn gave out free cups of water at a county fair. She had $2\frac{1}{2}$ gallons of water left after three hours. During the first hour, she passed out $3\frac{1}{2}$ gallons of water. She passed out $1\frac{1}{2}$ gallons of water in each of the second and third hours. How much water did Lynn start with?

Answer ______________________________

Evaluate Explain how you could check your answer by working forward from your answer.

____

____

7. At a block party, 3 friends take turns jumping over a bar. The bar is moved up $6\frac{3}{4}$ inches after each round—after all 3 friends jump. The third friend won in the fifth round of jumping. She jumped $34\frac{1}{2}$ inches after her friends missed at that height. What was the height of the bar for the first round of jumping?

Answer ______________________________

Discuss Does the answer change if there are 5 friends and all but 1 miss the bar on the fifth jump? Explain your thinking.

____

____

Choose one of the problems in this lesson. Change two things about the mathematics in the problem to create a new problem. Write and solve this new problem.

Strategy Focus

Use Logical Reasoning

MATH FOCUS: Decimal Concepts

Learn About It

Read the Problem

Adam has 4 pets. Their ages are 2 years, 2.25 years, 2.75 years, and 2.5 years. His dog is the youngest pet. His cat is the oldest pet. His bird is older than his mouse. How old is Adam's bird?

Reread Ask yourself questions as you read the problem.

- What is this problem about?

____

- What kind of information is given in the problem?

____

- What do I have to find?

____

Mark the Text

Search for Information

Read the problem again. Circle the numbers and words you need to answer the question.

Record Write what you learned about Adam's pets and their ages.

Adam has ________ pets.

His dog is the ____________ pet.

His cat is the ____________ pet.

His ____________ is older than his ____________.

You can choose a strategy to help you organize this information.

Decide What to Do

You know the kinds of pets Adam has. You know their ages. You need to match the ages and the pets.

Ask How can I find out how old Adam's bird is?

- The strategy *Use Logical Reasoning* can help me.
- I can organize the clues in a chart.

When you think logically, you may actually know more than you think. For example, if you know an animal is a dog, then you know it is *not* a cat.

Use Your Ideas

Step 1 Use the first clue: *His dog is the youngest pet.* Put a ✔ in the box to show this. No other pet is 2, so put an X in the empty boxes in that column. The dog is not any of the other ages. So put an X in the empty boxes in the Dog row.

	2	2.25	2.5	2.75
Bird	X			
Cat	X			✔
Dog	✔	X	X	X
Mouse	X			

Step 2 Use the second clue: *His cat is the oldest pet.* Use a ✔ to show this. Put an X in the empty boxes in the Cat row and in the 2.75 column.

Step 3 Use the third clue: *His bird is older than his mouse.* Which number is greater: 2.25 or 2.5? ________

Complete the chart.

So Adam's bird is ________ years old.

Review Your Work

Did you remember to put an X in each empty box of the column and the row in which you put a ✔?

Explain Can you start with the third clue? Why or why not?

____

____

Try It

Solve the problem.

Ask Yourself

How can I organize the information given in the clues?

① Joe gives these clues that tell his age.

A. I can write my age as a 3-digit decimal that is greater than 8 and less than 10.
B. The hundredths digit is 4 less than the ones digit.
C. The sum of all the digits is 16.
D. The tenths digit does not equal the hundredths digit.

How old is Joe?

Read the Problem and Search for Information

Find the question you need to answer. Reread the problem to be sure you understand the information.

Decide What to Do and Use Your Ideas

You can use logical reasoning to organize information in a place-value chart. Work through the clues.

Step 1 You know that either the digit 8 or 9 can go in the ones place.

Ones		Tenths	Hundredths
8	.		
9	.		

Step 2 Write the hundredths digit for each ones digit.

Step 3 Write the tenths digit for each number.

Step 4 Is the tenths digit of 8.44 equal to its hundredths digit? ________ Is the tenths digit of 9.25 equal to its hundredths digit? ________

So Joe's age is ________ years.

Review Your Work

Go back and check that your answer works for all four clues.

Describe How does the place-value chart help you find the correct decimal?

____

____

Apply Your Skills

Solve the problems.

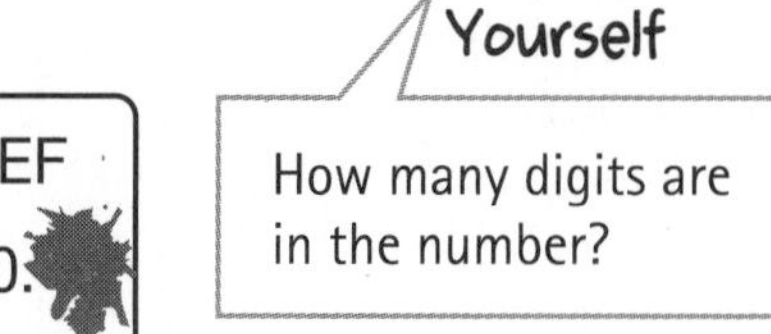

2 Rowan is at the library. She uses a number to help her look for a biography. She only knows the book's first three digits, which you can see at the right. Two digits are hidden by an ink blot. The hundredths digit is the same as the tens digit. The tenths digit is one less than the hundredths digit. What is the number on the book?

REF
920.

Hundreds	Tens	Ones		Tenths	Hundredths
9	2	0	.		

Hint Use a place-value chart to help you find the missing digits.

Answer ______________________________

Identify How did you know how many digits you need to find?

____

____

Ask Yourself

How can I write 15 minutes as a decimal?

3 Jaya, Fred, Gil, and Hasad sleep 7.75, 8.0, 8.25, and 8.5 hours each night, but not in that order. Fred sleeps fewer than 8 hours. Jaya and Hasad each sleep for more than 8 hours. Jaya sleeps 15 minutes less than Hasad. Put the four people in order from fewest to most hours slept.

Hint 15 minutes is $\frac{1}{4}$ of an hour.

	7.75	8.0	8.25	8.5
Jaya				
Fred				
Gil				
Hasad				

Answer ______________________________

Outline Write a brief description of how you could solve this problem without using a table.

____

____

Ask Yourself

Which clue should I use first?

4 Marcia runs the 50-meter dash in under 11 seconds. She gives these clues for her time.

A. My time is a 4-digit decimal that is greater than 10.
B. The tenths digit is 7 more than the ones digit.
C. The difference between the tenths and hundredths digits is 2.
D. The sum of all the digits is 13.

What is Marcia's time?

Hint The four digits you need to find are the tens, ones, tenths, and hundredths.

Tens	Ones		Tenths	Hundredths
		.		

Answer ______________________________

Sequence What steps did you use to find the correct answer?

____

____

Hint There are five friends and five distances in this problem.

5 Ed, Frank, Guy, Kai, and Cindy live on the same street as the school they attend. They live 0.3 kilometer, 1.3 kilometers, 1.4 kilometers, 0.7 kilometer, and 1.1 kilometers from school. The four boys all walk past Cindy's house to get to school. Frank, Guy, and Ed live more than a kilometer from school. Ed lives 0.1 kilometer closer to school than Guy. Write the names of the students in order from who lives closest to school to who lives the farthest away.

Ask Yourself

Will putting the distances in order help you solve this problem?

Answer ______________________________

Choose What is another strategy you could use to solve this problem? Explain.

____

____

On Your Own

Solve the problems. Show your work.

6. The softball team bought a new juice cooler. The amount of juice it holds is a 4-digit number. It holds more than 10 liters but less than 12 liters of juice. The tenths digit is half of the hundredths digit. The hundredths digit is 5 more than the tens digit. The sum of the digits is 11. How much juice can the cooler hold?

Answer ______________________________

Analyze Tom said that the cooler can hold 10.46 liters. What clue did he forget to use?

____

____

7. Five friends make meals that use different amounts of rice: $\frac{1}{2}$ cup, $\frac{3}{4}$ cup, $1\frac{1}{4}$ cups, $1\frac{1}{2}$ cups, and $2\frac{1}{4}$ cups. Henry uses $1\frac{1}{4}$ cups. Marta does not use the most or the least rice. The amount Irene uses is not $\frac{3}{4}$ cup or two times that amount. Kirk uses the most rice. Leah uses twice as much rice as Marta. How much rice does Irene use?

Answer ______________________________

Conclude How does knowing how to find equivalent fractions help solve this problem?

____

____

Choose one of the problems in this lesson. Change two things about the mathematics in the problem to create a new problem. Write and solve this new problem.

Lesson 12

Strategy Focus

Look for a Pattern

MATH FOCUS: Adding and Subtracting Decimals

Learn About It

Read the Problem

Train passengers can buy one-way and round-trip tickets to 5 travel zones. There is a pattern in the way the ticket prices change from zone to zone. The sign that gives the prices has been torn. What is the price of a one-way ticket to Zone 5?

	Zone 1	Zone 2	Zone 3	Zone 4	Zo
One-Way Ticket	$2.50	$3.75	$5.00	$6.25	
Round-Trip Ticket	$4.75	$7.25	$9.75	$	

Reread Be sure you understand the problem and the sign.

- What types of tickets are for sale?

____

- What does the problem tell me about the ticket costs?

____

- What is the question I need to answer?

____

Search for Information

Read the problem and the sign again. Some information may not be needed to solve the problem.

Record Write what you know about the ticket prices.

The cost of a one-way ticket to Zone 1 is ________ .

The cost of a one-way ticket to Zone 2 is ________ .

The cost of a one-way ticket to Zone 3 is ________ .

The cost of a one-way ticket to Zone 4 is ________ .

You can find a way to solve this problem by looking carefully at this information.

Decide What to Do

You know the ticket prices for 4 zones. You know that there is a pattern in the way the ticket prices change.

Ask How can I find the price of a one-way ticket to Zone 5?

- I can make a table to organize the information.
- Then I can use the strategy *Look for a Pattern*.

Use Your Ideas

	Zone 1	Zone 2	Zone 3	Zone 4	Zone 5
One-Way Ticket	\$2.50	\$3.75	\$5.00	\$6.25	

Change: _____ _____ _____ _____

Step 1 Find the change in price between Zones 1 and 2.

What amount do you add to \$2.50 to get to \$3.75?

\$2.50 + _______ = \$3.75

Write the changes in price under the arrows between the zones.

Step 2 Find the change in price from Zone 2 to Zone 3. Find the change in price between Zones 3 and 4. In each case, the increase is the same. The rule is add _______ .

Step 3 Use the rule to find the cost to Zone 5.

\$6.25 + _______ = _______

So a one-way ticket to Zone 5 costs _______

Review Your Work

Check that you did not make any computation errors.

Summarize Write how you found the pattern.

____

____

Try It

Solve the problem.

1. People who return overdue DVDs to the library must pay a fine. The amount of the fine increases each day. If the pattern continues, what is the fine for a DVD that is 6 days late?

NO FINE FOR 1 DAY LATE!
2 Days Late: $0.75
3 Days Late: $1.75
4 Days Late: $3.00
5 Days Late: $4.50

Read the Problem and Search for Information

Restate the question. Identify the information you need.

Decide What to Do and Use Your Ideas

You can use the strategy *Look for a Pattern.*

Step 1 Record the fines and the increase from one day to the next. Look for a pattern in the change in increase.

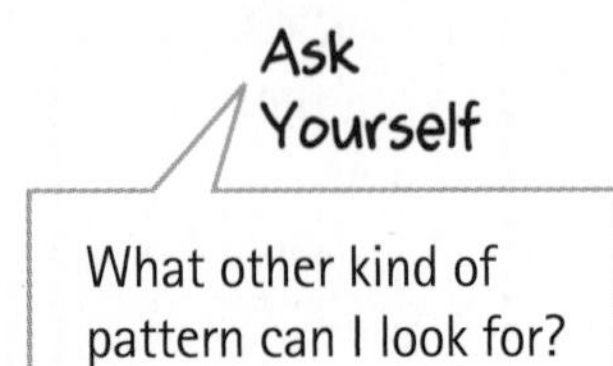

Days Late	1	2	3	4	5	6
Fine	None	$0.75	$1.75	$3.00	$4.50	?

Increase: + $0.75 + $1.00 + $1.25 + $1.50 + ?

Change in Increase: + $0.25 + $0.25 + $0.25 ______

Step 2 The change in increase is the same.

$1.50 + $0.25 = ________

Step 3 Add that amount to the fine for Day 5.

$4.50 + ________ = ________

The fine for a DVD that is 6 days late is ________ .

Review Your Work

Check that you lined up the decimal points correctly.

Contrast How is the pattern in this problem different from the pattern in the problem in Learn About It?

____

____

Apply Your Skills

Solve the problems.

2. At a sandwich shop, long sandwiches called subs are sold by the inch. Mike wants a 15-inch sub. If the pattern continues, how much will a 15-inch tuna sub cost?

Subs

Size	3 inches	6 inches	9 inches	12 inches
Price	$3.50	$6.00	$8.50	$11.00

Change: ______ ______ ______

Answer ____________________

Ask Yourself
What amount do I add to $3.50 to get $6.00?

Hint Use the change in price to find how much it will cost for the next size sub.

Conclude How does making a table help you find a rule for the pattern?

____

____

3. An office supply store sells notebooks. They are 10 inches long and 8 inches wide. The store sets the price by how many pages the notebook has. One of the workers started to make a sign for the notebooks, but did not finish it. Find the prices of the 125-page and 150-page notebooks.

Number of Pages	25	50	75	100	125	150
Price	72¢	90¢	$1.08	$1.26		

Change: ______ ______ ______ ______ ______

Answer ____________________

Ask Yourself
How can I change money amounts in cents to money amounts with dollar signs and decimal points?

Identify What information given in this problem is not needed to solve the problem?

Hint Find the rule for the prices.

____

____

Ask Yourself

How can I find a missing number in a pattern?

Hint Look for a pattern before and after the missing number.

4 Jo wants to buy some ribbon. Someone has scribbled on the sign that lists the prices. Jo wants to buy 3 feet of ribbon. How much will the ribbon cost?

Length (ft)	6	5	4	3	2	1
Cost ($)	3.90	3.25	2.60	[illegible]	1.30	[illegible]

Decrease: ______ ______ ______ ______ ______

Answer ____________________

Compare How is the pattern in this problem different from the patterns you have found in other problems?

____

____

Hint Check to see if the increase in cost from one quantity to the next is the same.

5 Mrs. Tang sells apples at a farmer's market. She charges different prices for different quantities. The prices are shown below. If the pattern continues, how much will 6 apples cost?

1 apple costs $0.75.

2 apples cost $1.45.

3 apples cost $2.10.

4 apples cost $2.70.

5 apples cost $3.25.

Ask Yourself

Since the prices do not increase by the same amount each time, what do I do next?

$0.75 → $1.45 → $2.10 → $2.70 → $3.25 → ______

+ ______ + ______ + ______ + ______ + ______

Answer ____________________

Analyze Mrs. Tang said that after 7 apples, each extra apple costs $0.40. Why would she say that?

____

____

On Your Own

Solve the problems. Show your work.

6. Mario rides his bike home from school. His home is 2.5 miles away from school. After 6 minutes, he is 2 miles from home. After 12 minutes, Mario is 1.5 miles away from home. If this pattern continues, how many minutes will it take Mario to get home?

Answer ______________________________

Sequence What steps did you follow to solve this problem?

____

____

7. At a juice bar, you can buy containers of juice in six sizes. The sign showing the prices is incomplete. If the pattern continues, how much will Bonnie pay for 1.5 liters of apple juice?

	0.25 Liter	0.5 Liter	0.75 Liter	1.0 Liter	1.25 Liters	1.5 Liters
Apple Juice	$0.95	$1.85	$2.70	$3.50		

Answer ______________________________

Elaborate What is the pattern in the apple juice prices?

____

____

Create

Choose one of the problems in this lesson. Change one or both patterns in the problem to create a new problem. Be sure that you still have patterns. Write and solve this new problem.

UNIT 3 Review What You Learned

In this unit, you worked with four problem-solving strategies. You can often use more than one strategy to solve a problem. So if a strategy does not seem to be working, try a different one.

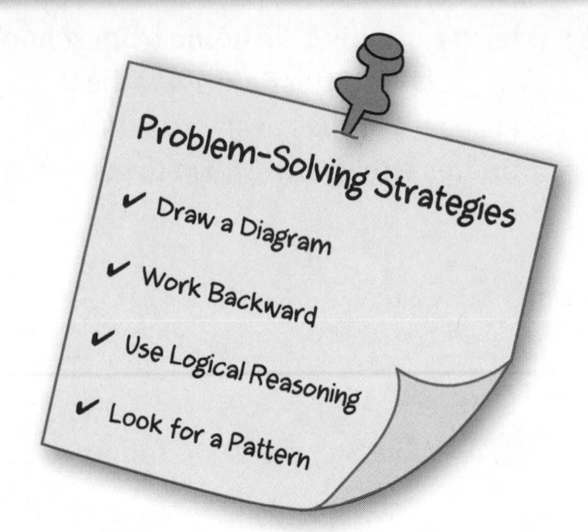

Solve each problem. Show your work. Record the strategy you use.

1. Alice has $6.75 after going shopping. She spent $13.45 at the grocery store. She also paid $3.00 for a magazine and $6.80 for a book. How much money did Alice have when she started shopping?

Answer ______________________

Strategy ______________________

2. Martha is planning to build a fence along one side of her back yard. The fence will be 20 feet long. It will have a fence post at one end and every 4 feet after that. Each fence post will be 3.5 feet tall. How much lumber will Martha need for the fence posts?

Answer ______________________

Strategy ______________________

3. Jon has 8 boxes. Each box is either all red, all blue, or both red and blue. One fourth of Jon's boxes are both red and blue. One half of his boxes are either all blue or part blue. What fraction of the boxes is all red?

Answer ______________________________

Strategy ______________________________

4. Reggie is drawing triangles. He wants to show a fraction pattern. He has shaded and labeled 4 of the 6 triangles in his pattern. What is the pattern of the fractions of the large triangle that are shaded? Show how he will shade and label the last 2 triangles.

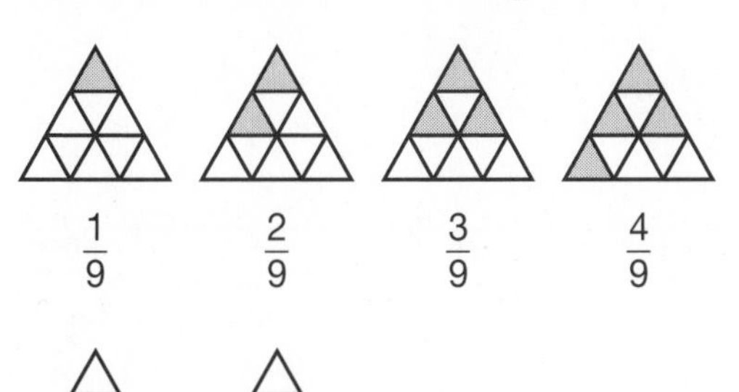

____ ____

Answer ______________________________

Strategy ______________________________

5. Wendy is thinking of a mixed number.

- The number is greater than 2 and less than 5.
- Its whole number part is $\frac{1}{2}$ of the denominator of the fraction part.
- Its numerator is $\frac{1}{2}$ of the sum of its whole number part and the denominator of the fraction part.
- Its fraction part is in simplest form.

What is the mixed number?

Answer ______________________________

Strategy ______________________________

Explain why your answer is the only possible answer to this problem.

Solve each problem. Show your work. Record the strategy you use.

6. Mia has $1.05 in her pocket in quarters and nickels. What is the least number of coins she might have? If she has the least number of coins, what are the coins? What is the greatest number of coins she might have? If she has the greatest number of coins, what are the coins?

Answer ______________________

Strategy ______________________

7. A chef made a cake for his diner. He served $\frac{1}{6}$ of the cake in the first hour. He served $\frac{5}{12}$ during the second hour. Then he served $\frac{1}{12}$ of the cake. How much of the cake was left? Write your answer in simplest form.

Answer ______________________

Strategy ______________________

8. This table shows how many points each problem on a math test is worth. The numbers in the *Points* row follow a pattern.

Problem	1	2	3	4	5	Total
Points	8.8	14.4		25.6	31.2	100

How many points is problem 3 worth?

Answer ______________________

Strategy ______________________

Explain how you could solve this problem without using a pattern.

9. Find and extend the pattern. Write the next two fractions. Describe the rule.

$\frac{1}{12}, \frac{1}{6}, \frac{1}{4}, \frac{1}{3}$, ________, ________

Answer ____________________

Strategy ____________________

10. Nick and his mom are making curtains for his room. After they made 2 curtains, they had $5\frac{3}{4}$ yards of fabric left. They used $1\frac{1}{2}$ yards of fabric for each curtain. How much fabric did they start with?

Answer ____________________

Strategy ____________________

Write About It

Look back at Problem 10. Describe how you used the information in the problem to choose a strategy for solving the problem.

____

____

Work Together: Plan a Scrapbook

Your soccer team is making a scrapbook as a gift for the coach. You will order photos to put on 6 pages. Your team has $22.00 to spend on the photos. How many photos will you order and in what sizes?

Photo Prices

Number of Photos	Size 5 in. × 7 in.	Size 8 in. × 10 in.
1	$1.75	$3.95
2	$3.45	$7.85
3	$5.10	$11.70
4	$6.70	$15.50

Plan

1. You must order some 5 in. × 7 in. photos and some 8 in. × 10 in. photos. Each page must have two 5 in. × 7 in. photos or one 8 in. × 10 in. photo.
2. The prices of the photos follow the patterns shown in the table above. Extend the table using the patterns you find.

Decide As a group, choose the number and size of photos to order.

Create Make a model to show the 6 pages.

Present As a group, share your decision and your model with the class. Include how much the photos will cost.

UNIT 4 Problem Solving Using Geometry

Unit Theme: Making Things

Did you know that geometry is all around us? Think about all the different shapes on the board games you play. Do you like to draw? Graphic artists use geometry in computer animations. In this unit, you will read about how math can be used in creative ways.

Math to Know

In this unit, you will use these math skills:

- Classify two- and three-dimensional figures
- Recognize and draw transformations
- Identify congruent and similar figures

Problem-Solving Strategies

- Draw a Diagram
- Make a Table
- Make a Graph

Link to the Theme

Finish the story. Write about the mural. Include some of the words from the list at the right.

Kiana and Theo passed a mural on their way home from school. They stopped to take a closer look. Kiana described what she liked about the mural.

____

____

____

____

____

____

____

Words to Use

shapes
triangles
circles
rectangles
sides

Use Math Language

Review Vocabulary

The list below shows vocabulary terms in this unit. Knowing the meaning of these terms will help you understand the problems.

horizontal	pyramid	rotation	translation
prism	reflection	transformation	vertical

Vocabulary Activity Word Pairs

Some math terms that are learned together can mean different things. Use terms from the above list to complete the following sentences.

1. Direction

a. A line that goes straight across from left to right is ________________ .

b. A line that goes straight up and down is ________________ .

2. Classifying Solid Figures

a. A certain solid figure has two faces that are congruent polygons. Its other faces are rectangles. This figure is a ________________ .

b. A certain solid figure has a base that is a polygon. Its other faces are triangles and they meet at one point. This figure is a ________________ .

Graphic Organizer Word Web

Complete the graphic organizer.

- In the top box, tell the meaning of *transformation*.
- In the middle row of boxes, write three different vocabulary terms related to *transformation*.
- In the bottom row of boxes, draw a picture to explain each of the three terms.

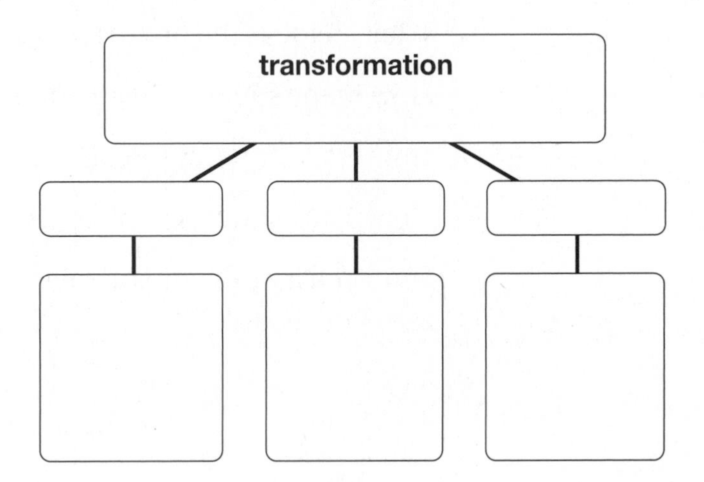

Strategy Focus

Draw a Diagram

MATH FOCUS: Polygons

Learn About It

Read the Problem

Cole is playing a game on a board made up of 30 squares. The game board is a rectangle that is 10 squares long and 3 squares wide. Players roll a number cube to find the number of spaces to move. Then they move their game pieces along the outside edge of the board. Cole moved his game piece once around the board. How many squares did he move?

Reread Ask yourself these questions as you read.

- What is the problem about?

____

- What am I being asked to do?

____

Mark the Text

Search for Information

Read the problem again. Look for important numbers.

Record Write what you know about the game board.

What shape is the board? ______________________________

How many squares make up the board? ____________

How many squares long is the board? ____________

How many squares wide is the board? ____________

You need a problem-solving strategy that can help you see the game board.

Decide What to Do

You know the shape of the game board. You know how many squares are on the board and how they are arranged.

Ask How can I find the number of squares along the edge of the board?

- I can use the strategy *Draw a Diagram.*
- I can use what I know to draw the rectangle. Then I can count the squares around the edge of the board.

A diagram can help you see what the game board looks like.

Use Your Ideas

Step 1 Use what you know to draw the game board.

Step 2 Count the squares along the outside edge of the board. You can write a number in each square to keep track.

There are ________ squares along the outside edge of the game board.

So Cole moved his game piece ________ squares.

Review Your Work

Check that the game board you drew matches the information in the problem.

Explain How does drawing a diagram help you answer the question?

____

____

Try It

Solve the problem.

① Li designed a game that uses a rectangle. The length of the rectangle is twice its width. The distance around the rectangle is 12 units. What is the width and the length of the rectangle?

Mark the Text

Read the Problem and Search for Information

Identify the shape of the figure and what you are asked to find.

Ask Yourself

What do I know about the width and the length of the rectangle?

Decide What to Do and Use Your Ideas

You can draw a diagram to solve the problem. You can guess the width and length. Then you can check and revise your answer.

Step 1 Suppose the rectangle's width is 1 unit.

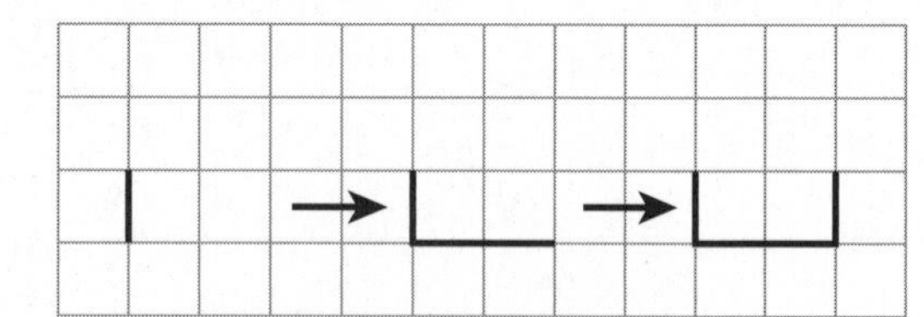

Draw a line segment that is 1 unit long. Since the length of the rectangle is twice its width, the length must be 2 units. Complete the rectangle.

The distance around this rectangle is ________ units.

Step 2 Draw a larger rectangle. Try a width of 2 units.

Draw a rectangle with a length that is twice its width.

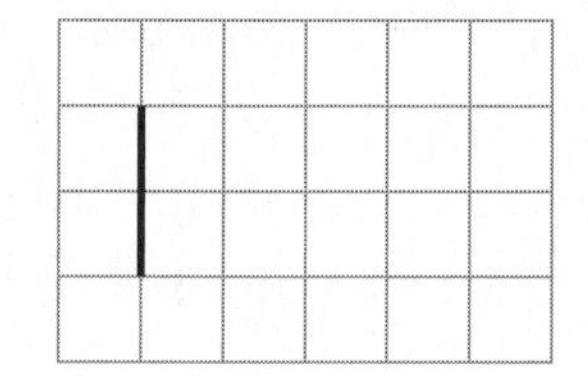

The distance around this rectangle is ________ units.

So the rectangle Li drew has a width of ________ units and a length of ________ units.

Review Your Work

Is the length of the second rectangle you drew twice its width?

Identify Why did you need to draw a larger rectangle after Step 1?

____

____

Apply Your Skills

Solve the problems.

(2) One kind of hopscotch game board is shown at the right. By combining shapes that are next to each other, how many different hexagons can you trace on this game board? Describe them.

10
9
7 8
6
5
4
2 3
1

Hint One of the shapes is part of each hexagon.

Ask Yourself

How many sides does a hexagon have?

Answer ______________________________

Recognize Which of the shapes can you combine to make a pentagon? Explain.

____

____

(3) Lamar is cutting out a hexagon, a triangle, an octagon, and a pentagon. Lamar cuts out the figures in this order: the one with fewest sides, the one with the most sides, the one with exactly 5 sides, and the last figure. Draw each shape. In what order did Lamar cut out the figures?

Hexagon	Octagon
Triangle	Pentagon

Ask Yourself

How many sides does each shape have?

Hint Write the number of sides each figure has next to each drawing.

Answer ______________________________

Analyze List the figures in order from fewest to most sides. Which two figures would complete a list of polygons up to 8 sides?

____

____

4 Kylie says that there is only one way to divide a square into 8 equal parts using exactly 4 lines. Max says there is more than one way. Who is correct? Include drawings to support your answer.

Hint Start with a 4 by 4 square.

Ask Yourself

What will the parts look like? Can they be squares? Rectangles? Triangles?

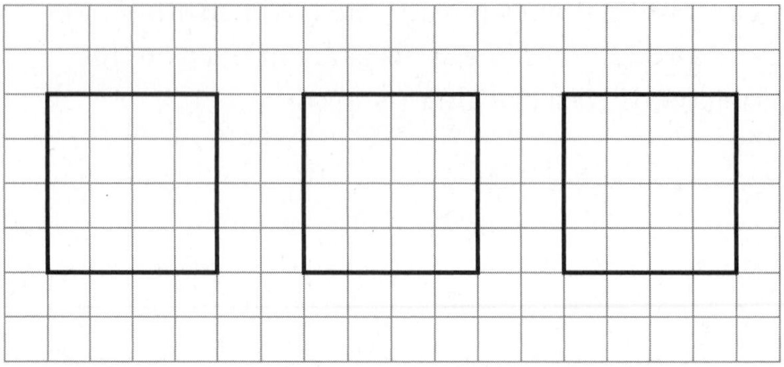

Answer ______________________________

Determine How does drawing the square on a grid help you to solve the problem?

____

____

5 Mel challenged Olivia to arrange 11 toothpicks to form 5 congruent triangles. Olivia solved the puzzle, as shown below. Then Olivia challenged Mel. She asked, "How many congruent triangles can you make with 15 toothpicks?" What should Mel answer?

Hint Number the triangles.

Ask Yourself

How many toothpicks do you need to add to the first triangle to form 2 triangles? 3 triangles?

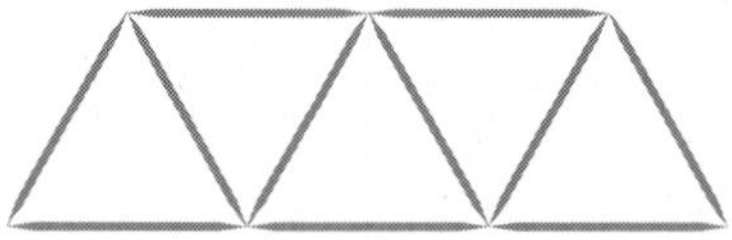

Answer ______________________________

Predict How many toothpicks would you need to form exactly 10 congruent triangles? Explain how you know.

____

____

On Your Own

Solve the problems. Show your work.

6. Krista has 4 sticks. Two sticks are 3 inches long and two sticks are 4 inches long. Describe two different figures Krista can make using 3 of the sticks. Describe one figure she can make using all 4 of the sticks.

Answer ______________________________

Modify Describe a different figure that Krista can make with all 4 sticks.

____

____

7. Maria and Vic are playing a game on a board made up of 9 squares. The game board is 3 squares long and 3 squares wide. Players place 6 triangles on the board so that there are exactly 2 triangles in each row and in each column. It is Maria's turn to play. Where might she put her 6 triangles?

Answer ______________________________

Adapt Suppose it is Vic's turn to play. Maria's triangles are still on the board. How can Vic move some of Maria's triangles so that there are exactly 2 triangles in each row and in each column?

____

____

Create

Make up a problem about designing a game board that can be solved using the strategy *Draw a Diagram*. Your game board should use triangles. Solve your problem.

Strategy Focus

Make a Table

MATH FOCUS: Solid Figures

Learn About It

Read the Problem

Len made models of a triangular prism and a triangular pyramid. Both solid figures have a triangle as a base. Which solid figure has more faces? Does the triangular prism have more edges, or more vertices? What about the triangular pyramid?

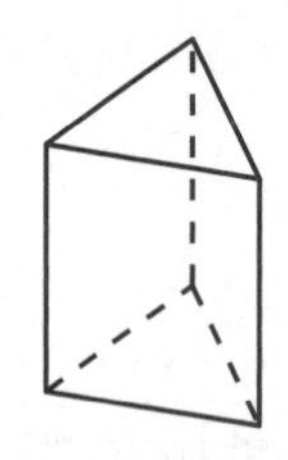

Triangular Prism

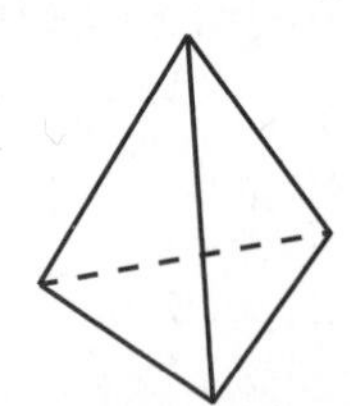

Triangular Pyramid

Reread Ask yourself these questions.

- What is the problem about?

____

- What questions am I asked to answer?

____

Search for Information

Read the problem again.

Record What do you need to find to answer the questions?

I need the number of ______________ in each solid figure.

I need the number of ___________________________ in the triangular prism.

I need the number of ___________________________ in the triangular pyramid.

Find a strategy that helps you use this information to solve the problem.

Decide What to Do

You have drawings of a triangular prism and a triangular pyramid. You need to know the numbers of faces, edges, and vertices of each solid figure. Then you need to compare the number of faces, edges, and vertices of each shape.

Ask How can I compare the numbers of faces, edges, and vertices?

- I can count the faces, edges, and vertices of each solid figure.
- I can use the strategy *Make a Table* to keep track of the results.

A table can help you organize the information.

Use Your Ideas

Step 1 Look at the drawings. Count and record the number of faces, edges, and vertices of each solid figure.

Solid Figure	Triangular Prism	Triangular Pyramid
Number of Faces		
Number of Edges		
Number of Vertices		

Step 2 Compare the number of faces.

A ______________________ has more faces.

Step 3 Compare the number of edges and vertices of each solid figure.

A triangular prism has more ____________ than ____________.

A triangular pyramid has more ____________ than ____________.

Review Your Work

You can draw on the drawings to check your results. Draw a dot at each vertex. Trace along each edge. Shade each of the faces.

Explain How does the table help you answer the question? What other questions can you answer using the table?

____

____

Try It

Solve the problem.

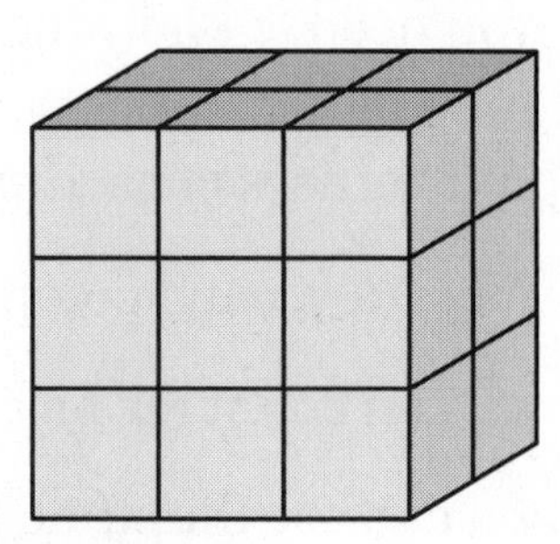

① Jen built a rectangular prism. She glued together 18 cubes of the same size. Matt picks up the prism and turns it around in his hands. He counts all the cube faces he can see. How many cube faces can he see?

Read the Problem and Search for Information

Think about the question you need to answer. Then picture the bottom, side, and back of the prism.

Decide What to Do and Use Your Ideas

You can use the strategy *Make a Table* to keep track of the information you collect.

As you work through the steps below, complete the table.

Ask Yourself

How many sides of Jen's prism does Matt need to look at?

Prism	Front	Back	Top	Bottom	Right Side	Left Side	Total
Cube Faces	9		6				

Step 1 Count and record the number of cube faces on each side of the rectangular prism.

Step 2 Find the sum of the number of faces of the cubes.

Matt can see a total of ________ faces of the cubes.

Review Your Work

Make sure you counted all the cube faces.

Visualize Suppose the prism is placed on a table in the same orientation as shown above. You can walk all around it but cannot pick it up. How many faces of the cubes would you be able to see? Tell how you know.

____

____

Apply Your Skills

Solve the problems.

② Sam made three pyramid models. One is square. One is pentagonal. One is hexagonal. Sam wants to see if there is a relationship between the number of faces and the number of vertices in a pyramid. Is there a relationship? If so, what is it?

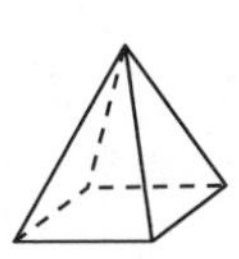 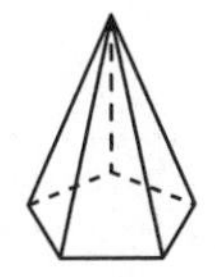 

Pyramid	Square	Pentagonal	Hexagonal
Number of Faces			
Number of Vertices			

Answer ______________________________

Hint You can draw on the diagrams to help you count.

Ask Yourself What is the pattern?

Compare Look at each solid figure again. How does the number of edges compare to the number of faces or vertices?

____

____

③ Nets for some solid figures are shown in the table. Suppose you use the nets to build models of the shapes. Complete the table. How are the shapes alike? How are they different?

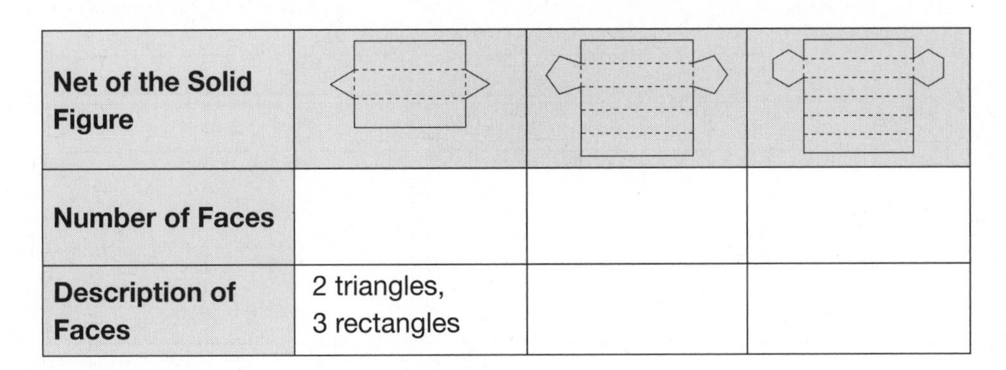

Net of the Solid Figure			
Number of Faces			
Description of Faces	2 triangles, 3 rectangles		

Answer ______________________________

Hint Think about cutting out and folding the figure along the dotted lines.

Ask Yourself What two-dimensional shapes are the faces of each prism?

Determine What is another question this problem could ask?

____

____

Hint The cubes form a rectangular prism. If you count the bottom edges of the pyramid, you do no need to count the top edges of the rectangular prism.

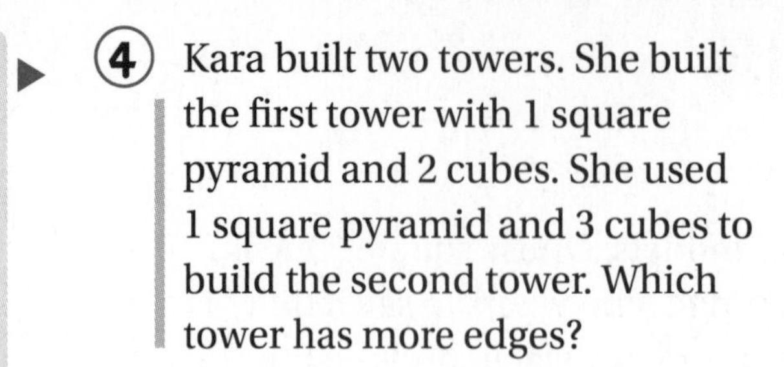

④ Kara built two towers. She built the first tower with 1 square pyramid and 2 cubes. She used 1 square pyramid and 3 cubes to build the second tower. Which tower has more edges?

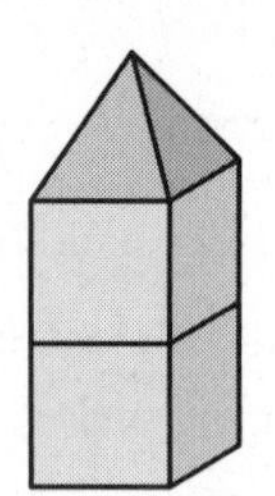

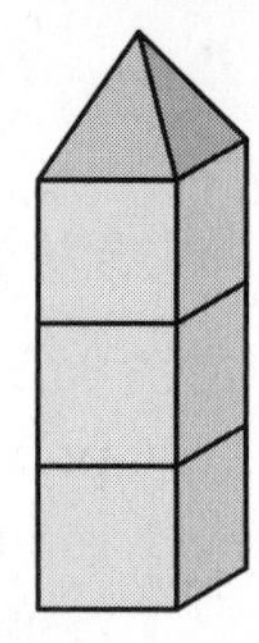

Ask Yourself

Do the pyramid and the rectangular prism share any edges?

Pyramid		
Bottom Edges of Rectangular Prism		
Vertical Edges of Rectangular Prism		

Answer ____________________

Analyze Would the number of edges change if Kara used 1 square pyramid and 5 cubes to build the tower? Explain.

____

____

Hint Choose two more prisms. Draw them to help you count the vertices, faces, and edges.

⑤ Jed knows all about solid figures. He knows that for all prisms, the expression below is always equal to the same number.

number of vertices + number of faces − number of edges

What is that number? Show at least 3 examples.

Ask Yourself

What will I label the rows and the columns in the table?

Cube	8	+	6	−		=	
		+		−		=	
		+		−		=	

Answer ____________________

Predict Will the number of vertices + number of faces − number of edges be the same for an octagonal prism as it is for a cube?

____

____

On Your Own

Solve the problems. Show your work.

6. Ana stacks cereal boxes side by side so that they touch each other. The boxes are on a display table at the front of the store. The stack is 5 boxes across and 3 boxes high. Ana needs to put a sticker on every side of each box she can see when she walks around the table. How many stickers will Ana need?

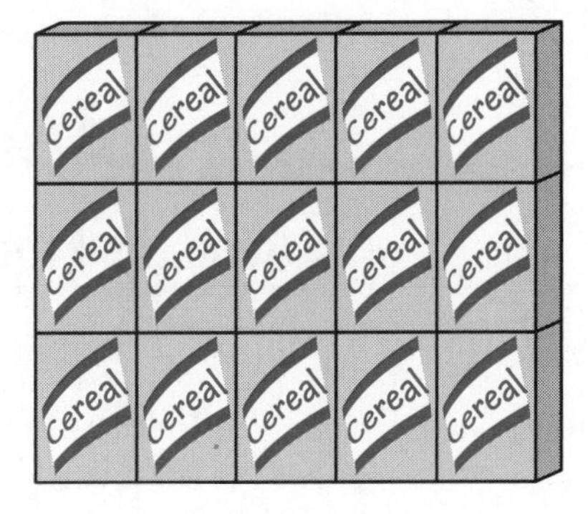

Answer ______________________________

Evaluate Suppose Chin gets an answer of 90 stickers. What did Chin do wrong?

____

____

7. Yuri wants to know if the expression below is always equal to the same number for all pyramids. If so, what is that number? If not, explain why not.

number of vertices + number of faces − number of edges

Answer ______________________________

Discuss What steps did you take to solve this problem?

____

____

Create

Make up a problem about using 3 solid figures to build a model house. Be sure your problem can be solved using the strategy *Make a Table*. Solve your problem.

Strategy Focus

Draw a Diagram

MATH FOCUS: Transformations and Symmetry

Learn About It

Read the Problem

Quinn is making a geometry puzzle. He finished one half of the design, which is shown below. He wants the puzzle to have a vertical line of symmetry that goes down the middle of the puzzle. What shapes will be in the finished puzzle?

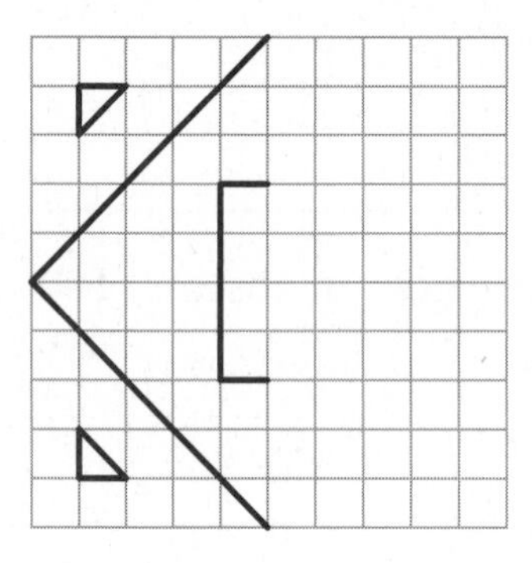

Reread Ask yourself these questions.

- What is Quinn making?

____

- What do I know about the puzzle?

____

- What do I need to find?

____

Search for Information

Read the problem again. Mark the information you need to answer the question.

Record Write what you know about the puzzle.

What words tell you how the puzzle design will look when it is finished?

____

You can use this information to choose a problem-solving strategy.

Decide What to Do

You know that Quinn is making a puzzle. You know that the puzzle has a vertical line of symmetry.

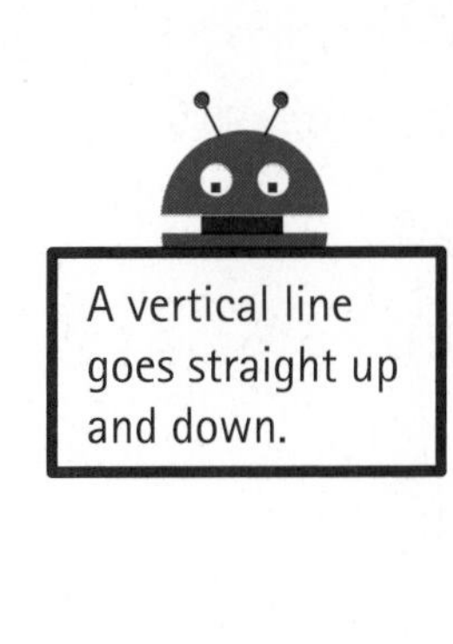

Ask How can I find the shapes that will be in the finished puzzle?

- I can use the strategy *Draw a Diagram.*
- I can use what I know about symmetry to finish the design. Then I can look at the shapes in my diagram.

Use Your Ideas

Step 1 Start with Quinn's design. Draw a vertical line in the middle of the puzzle to show the line of symmetry.

Step 2 Finish the design. Think about folding the puzzle along the line of symmetry. The part you draw should match Quinn's part.

Step 3 Look at the finished design. What shapes do you see?

The shapes in the puzzle are 1 ____________, 1 ____________, and 4 ____________.

Review Your Work

Trace the completed puzzle and fold your tracing to see if the parts match.

Explain Does the puzzle also have a horizontal line of symmetry? Tell how you know.

____

____

Try It

Solve the problem.

① Some geometry puzzle pieces are shaped like regular polygons. All the sides of a regular polygon are the same length. All its angles have the same measure. For this kind of polygon, how is the number of sides related to the number of lines of symmetry?

Read the Problem and Search for Information

Reread the problem. Underline the question the problem asks.

Decide What to Do and Use Your Ideas

You can use the strategy *Draw a Diagram.*

If I fold the triangle along each of the lines, will the parts match exactly?

Step 1 Draw regular polygons with 3, 4, 5, and 6 sides. Then draw all the lines of symmetry for each polygon.

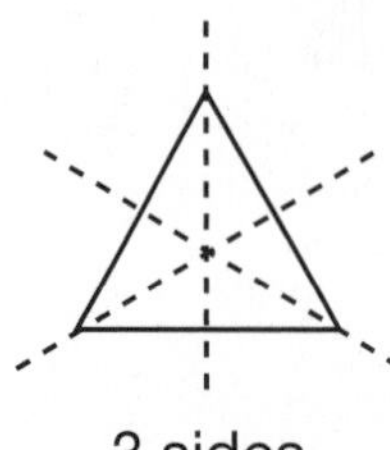

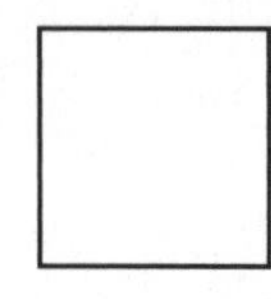

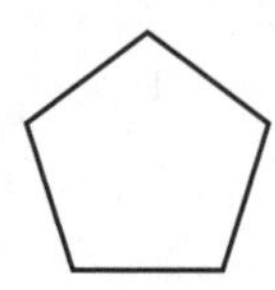

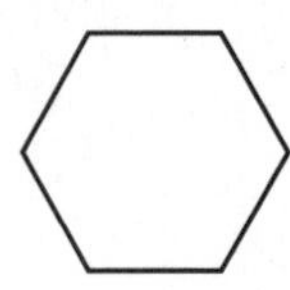

3 sides | 4 sides | 5 sides | 6 sides

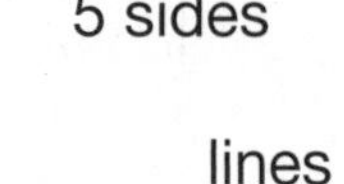

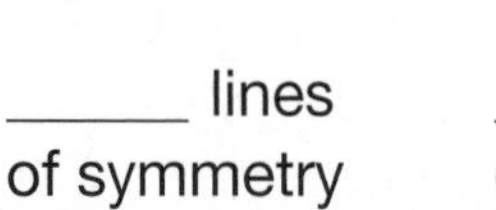

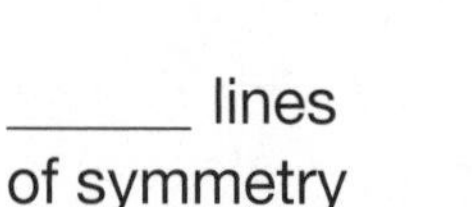

______ lines of symmetry | ______ lines of symmetry | ______ lines of symmetry | ______ lines of symmetry

Step 2 Count the lines of symmetry for each polygon. Write the number of lines of symmetry below the figure.

So for regular polygons, the number of lines of symmetry and the number of sides are ______________ .

Review Your Work

Check that you have drawn *all* possible lines of symmetry.

Predict How many lines of symmetry will a regular octagon have? Explain how you know.

____

____

Apply Your Skills

Solve the problems.

Ask Yourself

How would you undo "8 units to the right"?

② Look at the figure in the lower right corner of the diagram. It started out in the upper left corner. First, it was reflected across the dashed horizontal line. Next, the figure was translated 8 units to the right. Where was the figure at the start? Draw it. Then describe the transformations you used.

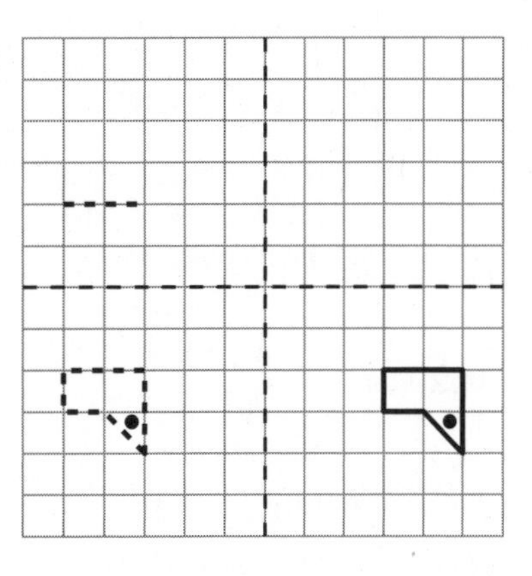

Hint You need to find where the figure was at the start. So you need to work backward.

Three types of transformations are translations (slides), reflections (flips), and rotations (turns).

Answer ____

Determine What is another way to get back to the first shape?

____

____

③ Which figure has *exactly* 2 lines of symmetry?

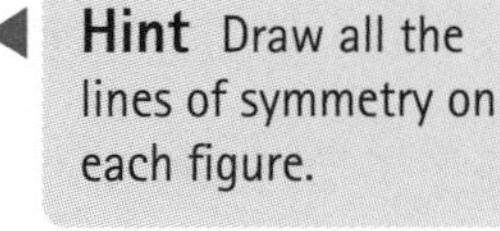

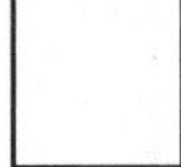
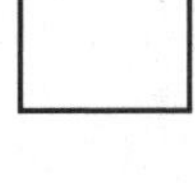

Ask Yourself

Can I fold the figure along exactly 2 lines so that the parts match exactly?

Answer ____

Analyze Suppose Jade's answer is the square. Is she correct? Why or why not?

____

____

Hint Figure A has been drawn for you on the grid. Use the grid to find transformations that give you one of the other figures.

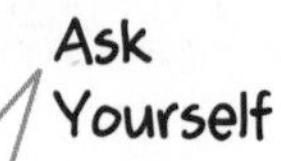

Which two figures are congruent, or match exactly?

④ Chen drew the three figures below. Which two figures are congruent? Use transformations to show how you know.

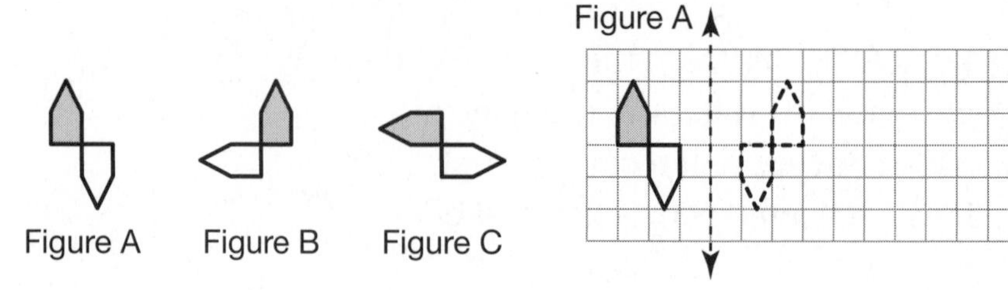

Answer ______________________________

Justify How could you convince someone that the two figures you chose are congruent?

____

____

Ask Yourself

Do I need to use more than one transformation?

Hint The triangle is facing a different direction.

⑤ Ali is putting together a tangram puzzle on her computer. She can reflect (flip), translate (slide), or rotate (turn) the pieces. What transformations can she use to fit the last piece in the puzzle?

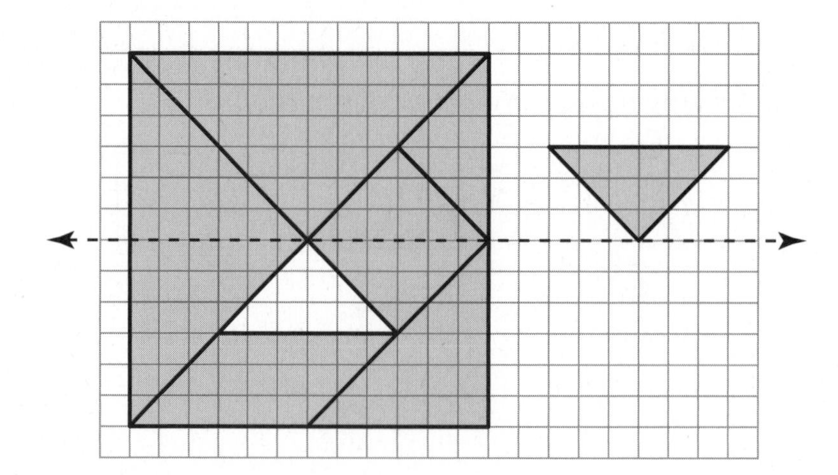

Answer ______________________________

Modify What is another way you can transform the piece to fit in the puzzle?

____

____

On Your Own

Solve the problems. Show your work.

6. Some capital letters have symmetry. The letter "V" has a vertical line of symmetry. One side is a reflection of the other. Kia wrote the note below in code. The letters have either a vertical or a horizontal line of symmetry. Some letters have both. What did her note say?

Answer ______________________________

Discuss How did you decide whether each letter had a vertical or a horizontal line of symmetry?

____

____

7. Suri creates a puzzle by translating hexagons as shown below. The middle hexagon is grey. Ring 1 of hexagons is white. Ring 2 is grey. Ring 3 is white, and so on. How many hexagons are in Ring 3?

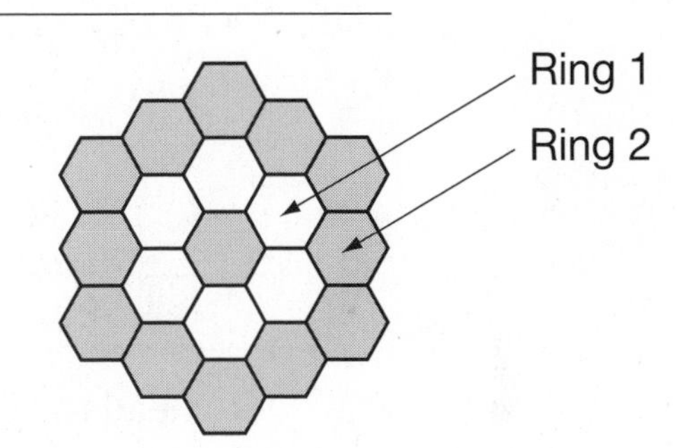

Answer ______________________________

Generalize What is the relationship between the number of the ring and the number of hexagons in the ring?

____

____

Write and solve a problem about drawing a design. It should have both a horizontal and a vertical line of symmetry. Make sure the problem can be solved using the strategy *Draw a Diagram*.

Strategy Focus
Make a Graph

MATH FOCUS: Shapes on the Coordinate Grid

Learn About It

Read the Problem

Patti is designing a banner. It will be in the shape of a quadrilateral. It will also have a vertical line of symmetry. She plots three vertices of the quadrilateral at (3, 7), (2, 3), and (12, 3). What are the coordinates of the fourth vertex? What kind of quadrilateral is Patti's banner?

Reread Use your own words to restate the problem.

- What is Patti designing?

____

- What do I know about the shape of the banner?

____

- What do I need to do to solve the problem?

____

Mark the Text

Search for Information

Read the problem again. Mark the important words in the problem. Check that you know the meaning of each one.

Record What information are you given in the problem?

The banner is in the shape of a ________________.

The shape has a ____________ line of symmetry.

The coordinates you know are ____________, ____________, and ____________.

A strategy that pictures the information can help you solve the problem.

Decide What to Do

You know that you can plot the points on a coordinate grid. You can use the line of symmetry to find the missing coordinates.

Ask How can I find the coordinates of the fourth vertex and the kind of quadrilateral Patti's banner is?

- I can use the strategy *Make a Graph* to plot the points on a coordinate grid.
- Next, I can draw a vertical line of symmetry and find the coordinates of the fourth vertex. Then I can connect the points and name the quadrilateral.

If you fold a shape along a line of symmetry, the parts will match exactly.

Use Your Ideas

Step 1 Plot the given points.

Step 2 The vertical line of symmetry will pass halfway between (2, 3) and (12, 3). Draw the vertical line of symmetry.

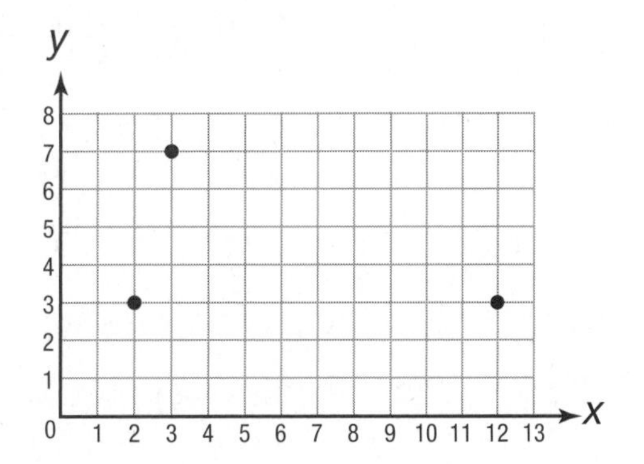

Step 3 Count the number of squares between (3, 7) and the line of symmetry. Plot a point the same distance away on the other side of the line of symmetry. The coordinates of that point are ______________ . Connect the points.

So Patti drew a __________________ with coordinates (3, 7), (2, 3), (12, 3), and ______________ .

Review Your Work

Check that you answered the questions that were asked.

Explain How does making a graph help you solve the problem?

____

____

Try It

Solve the problem.

① Hector is drawing a letter of the alphabet. He uses these coordinates. He connects these points in order.

(3, 2) (3, 6) (5, 2)

One point is missing. Find the coordinates of the missing point. What letter is Hector drawing?

Read the Problem and Search for Information

Reread to make sense of what the problem is about. Identify the information you need to solve the problem.

Decide What to Do and Use Your Ideas

You know that Hector is drawing a letter. You also know the ________________ he used to draw the letter so far.

You can use the strategy *Make a Graph* to solve the problem.

To plot a point, which way do I move first—to the right, or up?

Step 1 Plot the points you know. Connect the points in order.

Step 2 Plot the missing point. Complete the letter.

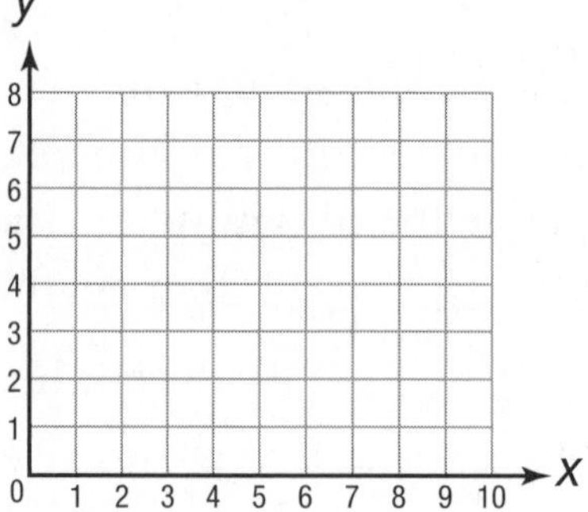

The coordinates of the missing point are ____________.
Hector is drawing the letter ________.

Review Your Work

Check that you plotted the points correctly.

Recognize Why is the information that Hector connects the points in order necessary? Explain.

____

____

Apply Your Skills

Solve the problems.

2 Alma wanted to draw two congruent rectangles. These are the coordinates of the vertices of the rectangles Alma drew.

Rectangle A (1, 1), (1, 4), (3, 4), (3, 1)

Rectangle B (5, 1), (5, 4), (9, 1), (9, 4)

Did Alma draw two congruent rectangles? Tell how you know.

Draw the second rectangle.

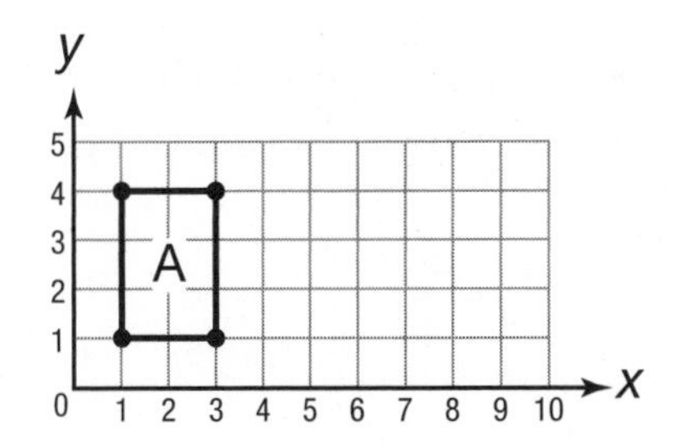

Answer ______________________________

Hint *Congruent* means exactly the same size and same shape.

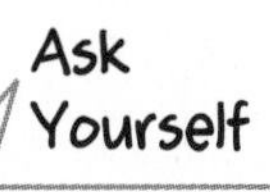

Do the rectangles match exactly?

Rearrange Explain how to change the coordinates of two vertices of Rectangle A so that the two rectangles are congruent.

____

____

3 Ray is drawing a design that has to be a right triangle. Two of the vertices of the triangle are at (2, 1) and (5, 4). At what coordinate could the other vertex of the right triangle be?

Hint A corner of a triangle is called a *vertex*.

Sketch a right triangle using the given points.

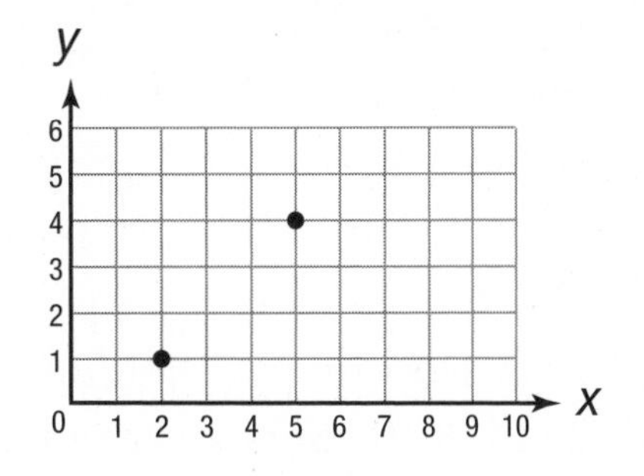

Ask Yourself

What must be true for a triangle to be a right triangle?

Answer ______________________________

Determine Ray says that there is more than one answer to this problem. Is he correct? Explain.

____

____

Hint Draw the two missing shapes.

Ask Yourself

Which corners of the missing shapes are the two missing points of the star?

④ Viv started a star quilt design. The design is made up of squares and right triangles. What are the coordinates of the two missing points of the star?

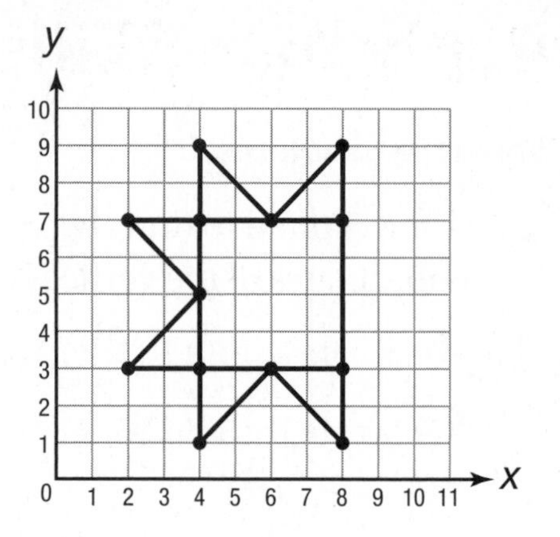

The two missing shapes are ________________ .

Answer ______________________________

Interpret How does seeing that some points are on the same vertical or horizontal line help you name the coordinates?

____

____

Hint There is more than one way to draw May's rectangle.

Ask Yourself

Did you draw a rectangle that is 2 units by 8 units?

⑤ Ethan and May are working on a mural. They draw two rectangles that are the same shape but not the same size. Ethan drew the rectangle shown. It measures 1 unit by 4 units. May drew a rectangle that is 2 units by 8 units. One vertex is at (8, 8) and another is at (8, 0). Draw May's rectangle. What could be the coordinates of the other 2 vertices?

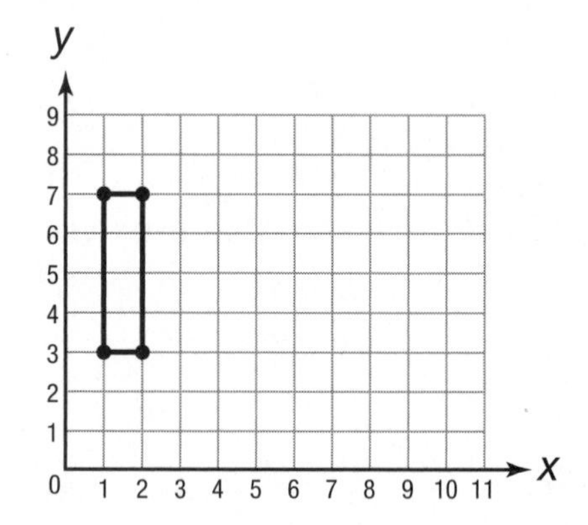

Answer ______________________________

Analyze How many different 2 by 8 rectangles with vertices at (8, 8) and (8, 0) could you draw on the above grid?

____

____

On Your Own

Solve the problems. Show your work.

⑥ Gail's driveway is 25 feet long and 10 feet wide. She decides to use chalk to draw a design on it. Gail starts by drawing a grid. Next, she draws points at (3, 4), (5, 7), (11, 7), and (9, 4). Then she connects the points in the order given above to form a four-sided shape. What shape did Gail draw?

Answer ______________________________

Judge What information is not necessary to solve the problem?

____

____

⑦ Leo was asked to slide this triangle 5 units to the right and then 4 units up. What are the coordinates of the triangle Leo should draw?

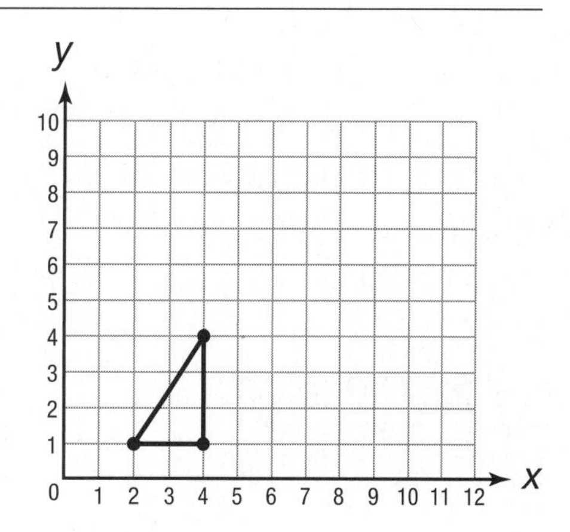

Answer ______________________________

Conclude What is another way to slide the triangle to the same position as Leo's triangle?

____

____

Create

Make up a problem about drawing squares on a coordinate grid. Be sure your problem can be solved using the strategy *Make a Graph*. Solve your problem.

Review What You Learned

In this unit, you worked with three problem-solving strategies. You can often use more than one strategy to solve a problem. So if a strategy does not seem to be working, try a different one.

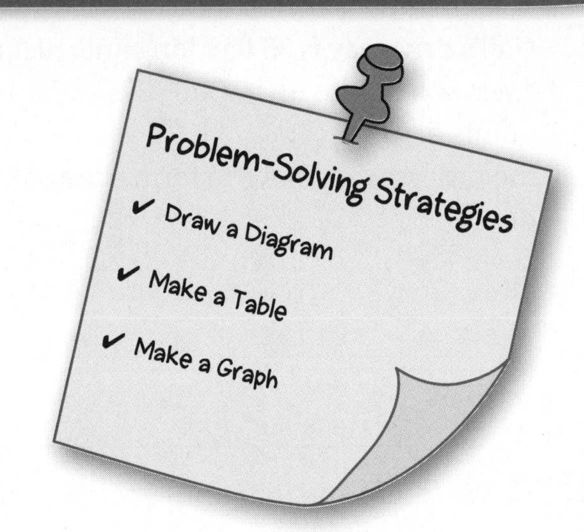

Solve each problem. Show your work. Record the strategy you use.

1. Mr. Cruz built a wall using 12 bricks. The wall is 3 bricks high and 4 bricks across. Then Mr. Cruz decided to paint the bricks. He painted all four sides of the wall and the top of the wall. How many faces of the bricks did he paint?

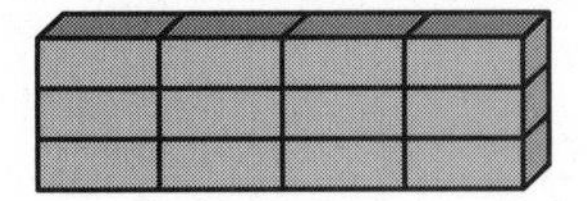

Answer ______________________

Strategy ______________________

2. Carla cuts a board that is 4 feet long into 6 equal pieces. Each cut takes 2 minutes. How long will it take Carla to make all of the cuts?

Answer ______________________

Strategy ______________________

3. Ben wants to draw a quadrilateral for a geometry game. The quadrilateral must have exactly one right angle. Draw a quadrilateral Ben could make. How many acute angles does it have?

Answer ______________________________

Strategy ______________________________

4. Look at the design below. Use transformations to describe how Figures A and B are related.

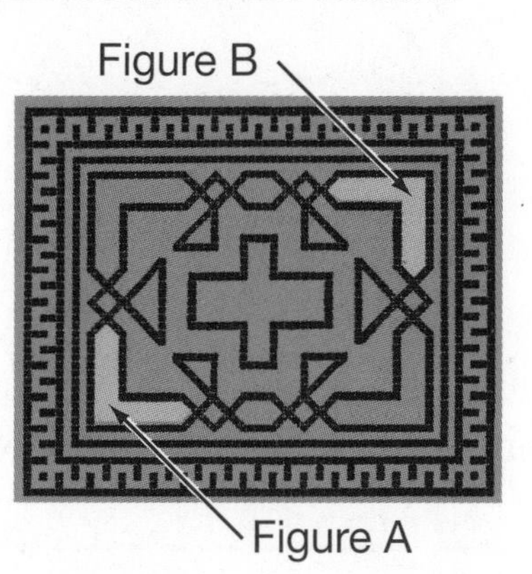

Answer ______________________________

Strategy ______________________________

5. Reflections of a square and an equilateral triangle look exactly the same as the original square and triangle. What are two other figures that have reflections that look exactly like the original figure?

Answer ______________________________

Strategy ______________________________

Explain your answer using words or drawings.

Solve each problem. Show your work. Record the strategy you use.

6. How many different squares are there in the diagram below? (Hint: The squares can be different sizes.)

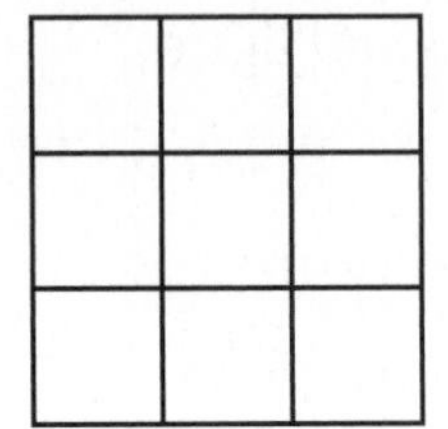

Answer ______________________

Strategy ______________________

7. Lea plots the points (4, 3), (6, 5), (6, 7), and (4, 8). She connects the points in that order with line segments. Then she draws a line segment to connect the last point to the first point. What shape has Lea drawn?

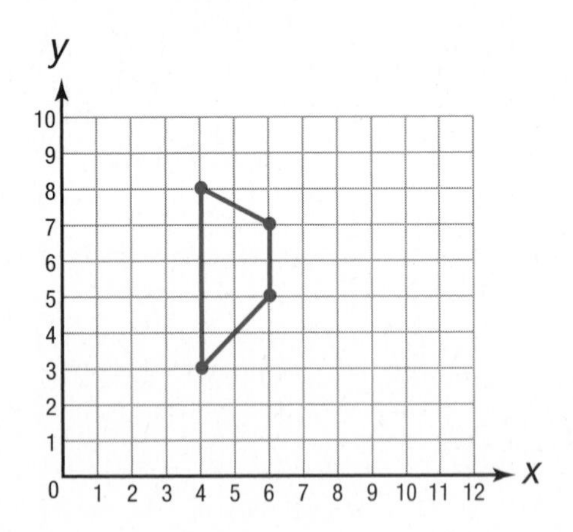

Answer ______________________

Strategy ______________________

8. Plot the points (0, 2), (5, 2), and (5, 7). Each coordinate is a vertex of a square. What are the coordinates for the fourth vertex of the square?

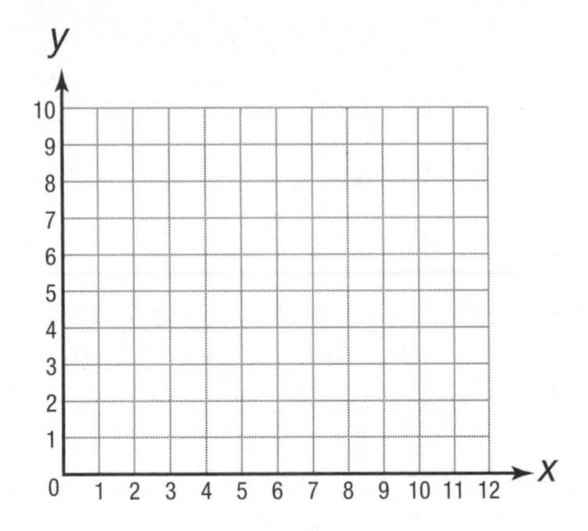

Answer ______________________

Strategy ______________________

Explain how you know that the figure is a square.

9. How many puzzle pieces in the size and shape of the small triangle will fit in the large triangle shown?

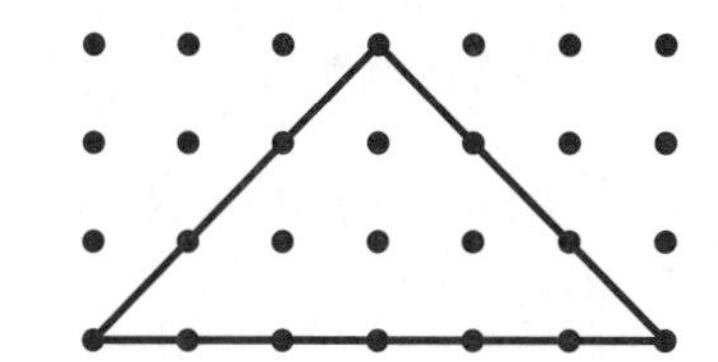

Answer ______________________________

Strategy ______________________________

10. How are a cube and a rectangular prism alike? How could they be different?

Answer ______________________________

Strategy ______________________________

Write About It

Look back at Problem 2. What information is not needed to solve the problem?

____

____

Work Together: Make a Flip Book

Your group will make flip books to show how animations work. When you flip the pages, the figure you drew looks like it is moving.

Plan

1. Use 10 or more small sheets of grid paper. Staple or put tape around the top edge to keep the pages together.
2. The first page you will see is the last page of your book. Start with a drawing on the last page and work backward.
3. Use very small transformations to move the figure on each page.

Decide Agree on a simple figure, such as the arrow that is shown. Each member of your group will make a flip book, using a different transformation to move that figure. Agree on which transformation you will use.

Design Create 10-page flip books.

Present Share your flip books with the class. Discuss the transformation you used and how it made the figure move.

UNIT 5 Problem Solving Using Measurement

Unit Theme:

Environments

No matter where you live, nature is close by. It may be in the form of a canyon, a forest, squirrels running through a city park, or even a tree growing from a sidewalk. In this unit, you will see how math is connected to Earth's many environments.

Math to Know

In this unit, you will use these math skills:

- Solve problems about time and temperature
- Compute using units in the customary and metric systems
- Use formulas for perimeter and area

Problem-Solving Strategies

- Work Backward
- Solve a Simpler Problem
- Look for a Pattern
- Write an Equation

Link to the Theme

Finish the story. Write about Chandra's calendar and her class trip. Use days, weeks, and months in your story.

Chandra looked at the calendar on her wall. She counted the days until her class trip to visit the nature center.

April

Sun.	Mon.	Tue.	Wed.	Thu.	Fri.	Sat.
	1	2	3	4	5	6
7	8	9	10	11	12	13
14	15	16	17	18	19	20
21	22	23	24	25	26	27
28	29	30				

Use Math Language

Review Vocabulary

The list below shows vocabulary terms in this unit. Knowing the meaning of these terms will help you understand the problems.

area	gram	meter	millimeter
centimeter	kilogram	milliliter	perimeter

Vocabulary Activity **Prefixes**

A *prefix* can provide a clue to a word's meaning. Use words from the list above to complete the following sentences.

1. The prefix *centi-* means "hundredth." A ________________ is $\frac{1}{100}$ of the length of a ________________ .
2. The prefix *kilo-* means "thousand." A ________________ is 1,000 times the mass of a ________________ .
3. The prefix *milli-* means "thousandth." A ________________ is $\frac{1}{1,000}$ of the length of a ________________ .
4. Another vocabulary word with the prefix *milli-* is ________________ .

Graphic Organizer **Word Diagram**

Complete the graphic organizer.

- Put these words in order based on the size of the unit, from least to greatest: *meter, centimeter,* and *millimeter.*
- Write the ordered words in the boxes.
- Explain how you knew how to order the words.

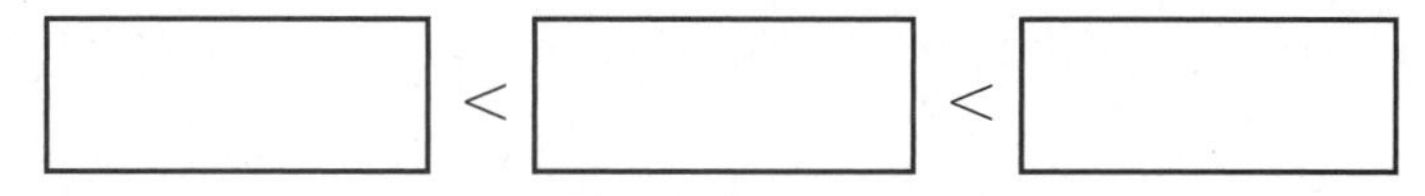

____

____

Strategy Focus
Work Backward

MATH FOCUS: Time and Temperature

Learn About It

Read the Problem

Ani's family rode mules into the Grand Canyon. It took 2 hours 30 minutes to get to the first stop at Indian Garden. They spent 45 minutes there eating lunch. Then they rode for another hour to Plateau Point. Ani knew that the group got to Plateau Point at 2:30 P.M. He forgot what time the ride started. What time did the mule ride start?

Reread Look for the descriptions of different parts of Ani's trip.

- In what places did Ani's family stop?

____

- What happened at 2:30 P.M.?

____

- What does the problem ask you to find out?

____

Mark the Text

Search for Information

As you read the problem again, look for the order of events on Ani's trip.

Record Write the events in order from first to last.

First, the mule ride started.

Next, ______________________________

Then ______________________________

Last, ______________________________

You will need a strategy that helps you connect these events to the times they happened.

Decide What to Do

You know the order of events for the mule ride. Now you need to use what you know to figure out when each event happened. You can look at a clock to help you.

Ask How can I find the start time for the trip?

- I can use the strategy *Work Backward* to find the start time.
- I know the time the mule ride ended. I know how long it took between events. So I can start from the end of the ride. Then I can subtract the time that passed between the events.

If you know the time at the end, you can work backward to the time at the start.

Use Your Ideas

Step 1 Start at the end of the mule ride. Work backward to find out what time lunch ended.

- Ani's family arrived at Plateau Point at ____________.
- How long did Ani's family ride after lunch to get to Plateau Point? ____________
- So lunch ended at ____________.

Step 2 Work backward until you get to the start of the trip.

- How long did Ani's family stay at Indian Garden? ____________
- Ani's family arrived at Indian Garden at ____________.
- How long did it take to get to Indian Garden?

  ____

So the trip began at ____________.

Review Your Work

Now start at the beginning and check all the times you found on the way to your answer. Make sure that the trip ends at 2:30 P.M.

Describe Why does it make sense to work backward to solve this problem?

____

____

Try It

Solve the problem.

① Wow! It is hot at the bottom of the Grand Canyon. A ranger told Cara that it is 24°F cooler at the South Rim. He also said that the temperature at the North Rim is 8°F cooler than it is at the South Rim. It is 70°F at the North Rim right now. What is the temperature at the bottom of the canyon?

Read the Problem and Search for Information

Think about the problem. Mark details that will help you.

Decide What to Do and Use Your Ideas

What operations should I use to find the temperature at the South Rim?

You can use the strategy *Work Backward*. Look back at the problem for facts about the temperatures and their differences.

Step 1 The temperature at the North Rim is 70°F. Find the temperature at the South Rim.

- The North Rim is ________ cooler than the South Rim.
- Add to find the temperature at the South Rim.

70°F + 8°F = ________

Step 2 Find the temperature at the bottom of the canyon.

- The South Rim is ________ cooler than the bottom.
- Add to find the temperature at the bottom.

78°F + 24°F = ________

So the temperature at the bottom is ________ .

Review Your Work

Work through the problem from the beginning using your answer.

Clarify Joan tells you that the word "cooler" means you should subtract. Is she right?

____

____

Apply Your Skills

Solve the problems.

② Molly is going on vacation in August. Her family is going to drive to the Grand Canyon. They will arrive 3 days after they leave. It will take the same amount of time to drive back. They will spend a week at the canyon. The family will get home on August 24. On what day will the family leave for vacation?

Ask Yourself

How many days are in a week?

Molly's family will arrive home on ____________ .

It will take them ________ days to drive home.

They will leave the Grand Canyon on ____________ .

They will stay ________ days at the Grand Canyon.

Their stay at the canyon will begin on ____________ .

It will take them ________ days to drive to the Grand Canyon.

Hint Work backward from August 24.

Answer ______________________________

Explain How could you check that your answer is correct?

____

____

③ On Monday, Dani began a five-day whitewater rafting trip down the Colorado River. She noticed that the temperature was 2°C warmer each morning during the trip. If the temperature was 28°C by Friday morning, what was the temperature on Monday morning?

Ask Yourself

Should I add or subtract to find the temperature on Monday morning?

Temperature on Friday: ________

Temperature on Thursday: ________

Temperature on Wednesday: ________

Hint You can count by 2s to solve this problem.

Answer ______________________________

Analyze Sonia tried to solve this problem by thinking, "Dani's trip lasted 5 days. The temperature was 2°C warmer each day. That makes a total of 10°C more." What mistake did Sonia make?

____

____

Ask Yourself

Does it take the family more or less than 2 hours altogether to set up camp and fix dinner?

Hint You can look at a clock or draw a clock face to help you.

④ When Dara's family goes camping, it takes them 1 hour and 15 minutes to set up camp. Then they make dinner, which takes about 30 minutes. If they want to eat dinner at 6:00 P.M., when do they have to start to set up camp?

Dara's family should start making dinner at ________ .

Will you add or subtract time to find out when they should start to set up camp? ____________

Answer ____

Compare If Dara's family spends a second night at the same campsite, what time would they have to start making dinner to eat at 6:00 P.M.?

____

____

Ask Yourself

Will my answer be greater than or less than 52?

Hint Use units for degrees in your answer.

⑤ Miko is watching the sun set at the Grand Canyon. She found a nice place to sit at 5:00 P.M. By 5:30 P.M., the temperature was 8°F cooler. Fifteen minutes later, the temperature had dropped 5°F to 52°F. It was chilly, but the sunset was beautiful! What was the temperature when Miko first sat down?

The temperature at the end is ________ .

The time at the end is ________ .

Answer ____

Interpret Miko started feeling cold when the temperature was about 55°F. At about what time did Miko start to feel cold?

____

____

On Your Own

Solve the problems. Show your work.

6. Oscar and Luis went hiking in the Grand Canyon. It was cool when they started. They saw Mooney Falls, which is about 200 feet high. When they stopped for lunch, it was 16°C warmer. By the time they finished their trip, it was 9°C cooler than at lunch. It was 25°C when they finished their trip. What was the temperature when they started?

Answer ______________________________

Identify What information is given that is not needed to solve the problem?

____

____

7. Kal and his family are planning when to go on their vacation. The last day of school is on a Friday. For the first two weeks after school gets out, Kal goes to science camp. On the Saturday after Kal's last day of camp, Kal's grandparents will come to stay for 7 days. Kal figured out that the first day they could leave for vacation would be Saturday, July 10. What is the date of Kal's last day of school?

June

Sun.	Mon.	Tue.	Wed.	Thu.	Fri.	Sat.
		1	2	3	4	5
6	7	8	9	10	11	12
13	14	15	16	17	18	19
20	21	22	23	24	25	26
27	28	29	30			

July

Sun.	Mon.	Tue.	Wed.	Thu.	Fri.	Sat.
				1	2	3
4	5	6	7	8	9	10
11	12	13	14	15	16	17
18	19	20	21	22	23	24
25	26	27	28	29	30	31

Answer ______________________________

Discuss How does the calendar help you solve the problem?

____

____

Create

Make up a problem that can be solved by using the strategy *Work Backward*. Your problem should be about temperatures that go up and down during the day. Solve your problem.

Strategy Focus

Solve a Simpler Problem

MATH FOCUS: Length, Weight, and Capacity Using the Customary System

Learn About It

Read the Problem

Rusty is the biggest horse on a farm. He weighs about 2,000 pounds! The height of a horse is often measured in *hands*. A hand is equal to 4 inches. Rusty is 19 hands high at the shoulder. How tall is Rusty in feet and inches?

Reread Think about what you read.

- What is the problem about?

____

- What measurements are given?

____

- What does the problem ask you to find out?

____

Mark the Text

Search for Information

Reread the problem. Mark the measurements and the measurement units used in the problem.

Record What information will help you solve the problem?

Rusty's height is ________ hands.

There are ________ inches in a hand.

When you want to solve a complicated problem, it can help to split it into smaller, simpler problems.

Decide What to Do

This problem asks you to change a measurement from one kind of unit to another. How do you decide what operation to use?

Ask How can I find Rusty's height in feet and inches?

- I can use the strategy *Solve a Simpler Problem* to see how to change hands to inches.
- Then I can see how to change from inches to feet and inches.

Use Your Ideas

Step 1 Change Rusty's height from hands to inches. First use simpler numbers to see what operation to use.

1 hand = ________ inches; 2 hands = ________ inches

To change from hands to inches, multiply by ________ .

Rusty is ________ hands tall.

19 × ____________ = ____________

Rusty is ________ inches tall.

Step 2 Change from inches to feet and inches. First use simpler numbers to see what operation to use.

12 inches = ________ foot; 24 inches = ________ feet

To change from inches to feet, ____________ by 12.

76 inches = ________ feet and ________ inches

When you divide by 12, the quotient is the number of feet and the remainder is the number of inches.

So Rusty is ______________ tall.

Review Your Work

Try working backward to check that your answer is equal to Rusty's height in hands.

Contrast Look at the operations you used to find the answer. How do they compare to the operations you used to check the answer?

____

____

Try It

Solve the problem.

(1) A miniature horse named Poco lives on a farm. Poco weighs less than 100 pounds. He drinks about 3 quarts of water a day. A larger horse named Bella drinks about 18 gallons of water a day. How many days would it take Poco to drink the 18 gallons that Bella drinks in 1 day?

Read the Problem and Search for Information

Determine what the problem is asking for. Identify and mark the measurements that will help you as you reread the problem.

Decide What to Do and Use Your Ideas

You can use the strategy *Solve a Simpler Problem* to break the problem into smaller, simpler pieces.

Ask Yourself

How can I convert from gallons to quarts?

Step 1 Decide how to change gallons to quarts. First, try simpler numbers.

1 gallon = 4 quarts 2 gallons = ________ quarts

So to change from gallons to quarts, I multiply by ________ .

The larger horse drinks ________ gallons each day.

18 gallons = ________ quarts

Step 2 Find the number of days it will take Poco to drink 72 quarts of water.

Poco drinks ________ quarts each day.

$72 \div 3 =$ ________

So it will take Poco ________ days to drink as much as Bella does in 1 day.

Review Your Work

Check that your answer makes sense.

Identify What information is given that is not needed?

____

____

Apply Your Skills

Solve the problems.

② Miniature horses are very short compared to other horses. They are about 30 inches tall to the top of the shoulder. A farmer's full-size horse is about 15 hands to the top of the shoulder. How many miniature horses would it take, standing on each other's shoulders, to reach the height of the farmer's horse?

Hint The height of miniature horses is given in a different unit than the height of the farmer's horse.

There are 4 inches in 1 hand, ________ inches in 2 hands, ________ inches in 3 hands, and ________ inches in 15 hands.

So the farmer's full-size horse is ________ inches tall.

1 miniature horse is only ________ inches tall.

So 30 inches × ________ = the height of the farmer's horse.

Ask Yourself

Should I change inches to hands or hands to inches?

Answer ________________________________

Explain Why must you first change the heights to the same units?

____

____

③ Mr. Lee works at a farm taking care of a horse. He gives the horse vitamin pellets. The pellets come in a 2-pound bucket. Each day, Mr. Lee gives 1 ounce of pellets to the horse. He wants to buy enough vitamin pellets to last 10 weeks. How many buckets should he buy?

Hint Break the problem into smaller steps.

1 week is ________ days, so 10 weeks = ________ days.

Mr. Lee gives the horse ________ ounce of vitamin pellets each day. So he will need ________ ounces for 10 weeks.

There are 16 ounces in 1 pound, so there are ________ ounces in 2 pounds.

Ask Yourself

How can I find out how many ounces are in a 2-pound bucket?

Answer ________________________________

Describe How do you use the relationship between the number of ounces and the number of buckets to solve the problem?

____

____

Ask Yourself

If $\frac{1}{2}$ cup of the vitamins lasts 1 day, how long will 1 cup last?

Hint There are 4 quarts in a gallon, 2 pints in a quart, and 2 cups in a pint.

④ Large horses sometimes have trouble with the joints in their legs and hips. To keep his horse healthy, Rico pours liquid vitamins over his horse's breakfast each day. Rico has 4 gallons of the vitamins. If the horse gets $\frac{1}{2}$ cup each day, how many days will 4 gallons last?

The horse needs ________ cup for 2 days.

1 gallon → ________ quarts → ________ pints → ________ cups.

Rico has ________ gallons, which is ________ cups.

Answer ____

Analyze Anna says that the 4 gallons will last 32 days. What error did she make?

____

____

Hint You can use simpler numbers to decide which operation to use to change between pounds to ounces.

Ask Yourself

Should I change ounces to pounds or pounds to ounces? Which will give me simpler numbers?

⑤ Mario helps out a local stable after school. The large horse there eats 18 pounds of grain and 40 pounds of hay every day. The miniature horse eats 12 ounces of grain and a few handfuls of hay each day. How many days would it take the miniature horse to eat all of the large horse's grain in daily meals of 12 ounces each?

The large horse eats ____________ of grain a day.

The miniature horse eats ____________ of grain a day.

1 pound = ________ ounces

Answer ____

Evaluate In this problem, did you change pounds to ounces or ounces to pounds? How did you decide which to do?

____

____

On Your Own

Solve the problems. Show your work.

6. A veterinarian wants Tina to add medicine to her horse's food each day. Tina will add the medicine for 12 days using the following schedule.

   $\frac{1}{4}$ cup each day for 4 days, $\frac{1}{2}$ cup each day for 4 more days, $\frac{3}{4}$ cup each day for the last 4 days

   Then the veterinarian will check on the horse. Will a 1-quart bottle of medicine last until the veterinarian comes back? Explain your answer.

Answer ______________________________

Predict How many days will a 1-quart bottle last? Explain.

____

____

7. Mr. Jones gives his 6 horses vitamins to keep them healthy. The vitamin pellets come in 50-pound bags. Mr. Jones gives each horse 8 ounces of the vitamins every day. Mr. Jones buys 20 bags of the vitamins. Will this last him a year? Explain your answer.

Answer ______________________________

Discuss How is this problem like Problem 3? How is it different?

____

____

Create

Create a problem about horses that involves measurements that are given in two different units. Be sure your problem can be solved using the strategy *Solve a Simpler Problem*.

Strategy Focus

Look for a Pattern

MATH FOCUS: Length, Mass, and Capacity Using the Metric System

Learn About It

Read the Problem

Ms. Arbor's class is hiking on a 600-meter trail in the forest. She ties yarn around tree branches to mark the trail. Every 20 meters from the start, she marks a tree with red yarn. Every 30 meters from the start, she marks a tree with yellow yarn. If a tree is marked with both red and yellow yarn, she also ties orange yarn around it. She uses 20 centimeters of yarn each time she marks a tree branch. How much orange yarn does Ms. Arbor use in all?

Reread Ask yourself these questions as you read the problem.

- What is Ms. Arbor doing?

____

- What is she using to mark the trail?

____

- How long is the trail?

____

- What does the problem ask you to find?

____

Mark the Text

Search for Information

Mark details in the problem that describe at what point each color of yarn is used. Circle any measurements in the problem.

Record How does Ms. Arbor mark the trees?

She uses red yarn to mark trees every ________ meters.

She uses yellow yarn to mark trees every ________ meters.

She uses orange yarn to mark trees that have ______________________________.

Think about what this information tells you.

Decide What to Do

The problem describes a pattern. It tells where each color of yarn is used. Use the information in the problem to figure out the pattern.

Ask How can I find out how much orange yarn Ms. Arbor uses?

- I can look for a pattern of trees marked with orange yarn.
- I can multiply that number of trees by the length of yarn needed for each tree. This will give me the total length of orange yarn needed.

Use Your Ideas

Step 1 Write a number pattern to show where Ms. Arbor ties red yarn and where she ties yellow yarn. Fill in the missing number of meters for each color from the start of the trail.

Red: 20, 40, 60, 80, 100, ________, ________, ________, ________,...

Yellow: 30, 60, 90, ________, ________, ________,...

Step 2 Look for numbers that are in both lists.

So the orange yarn pattern is 60, 120, and ________.

Step 3 Find the number of trees marked with orange yarn.

Continue the pattern for orange yarn up to 600.

60, 120, 180, 240, 300, 360, 420, ________, ________, ________,...

So ________ trees are marked with orange yarn.

Step 4 Find how much orange yarn Ms. Arbor uses.

Ms. Arbor uses ________ centimeters of yarn for each tree.

Multiply. $10 \times 20 =$ ________

So Ms. Arbor uses ________ centimeters of orange yarn.

The trees that are marked with both red and yellow yarn are also marked with orange yarn.

Review Your Work

Read through the problem again. Does your answer make sense?

Describe How did the patterns help you solve the problem?

____

____

Try It

Solve the problem.

① Keesha's cat Max climbs trees to chase squirrels. Max's mass is greater than a squirrel's. Today, Max crept out on a branch and it broke. Keesha wondered how many squirrels it would take to break the branch in the same way. Max's mass is 4 kilograms. An average squirrel has a mass of 700 grams. What is the least number of squirrels that would be greater than Max's mass?

Read the Problem and Search for Information

Think about the situation in the problem. Find out what the problem is asking you to answer. Then mark the details.

Decide What to Do and Use Your Ideas

You can look for a pattern in the number of squirrels and their total mass.

Ask Yourself

Should I change kilograms to grams or grams to kilograms?

Step 1 Change Max's mass from kilograms to grams.

Max has a mass of ________ kilograms.

There are 1,000 grams in 1 kilogram.

So Max has a mass of ________ grams.

Step 2 You can make a table to show the pattern.

Each squirrel adds 700 grams. Continue the pattern.

Number of Squirrels	1	2	3	4		
Mass (grams)	700	1,400	2,100	2,800		

So the least number of squirrels with a mass that is greater than Max's is ________ .

Review Your Work

Check that you answered the question that was asked.

Conclude How do you know that your answer makes sense?

____

____

Apply Your Skills

Solve the problems.

② Oscar brings 2 bottles of water on a hike with his family. Each bottle holds 1 liter of water. He drinks half of the water he has left at each stop on the hike. How many milliliters of water does Oscar have left after his fourth stop?

Stop Number	0	1	2	3	4
Water Left After Drinking (milliliters)	2,000				

Answer ______________________________

Determine Could you just divide 2,000 milliliters by 4 to find the answer? Why or why not?

____

____

Ask Yourself

How can I find how much water he drinks at each stop?

Hint Oscar begins the hike with 2,000 milliliters of water.

③ Mr. Bradley is training for a long hike. He plans to hike across two states. This week, he trains for five days. On the first day, he hikes 2 miles. On each of the next four days, he hikes 3 more miles than he did the day before. How many miles does Mr. Bradley hike this week?

Number of miles hiked each day: 2, 5, ________, ________, ________

Answer ______________________________

Analyze What information is not needed to solve the problem?

____

____

Hint The problem asks for the total number of miles hiked, not the number of miles hiked on the fifth day.

Ask Yourself

What rule should I use?

Ask Yourself

How many millimeters are in 40 centimeters?

Hint Use multiples of 5 for the number of ants to make a pattern.

④ Zak sees some ants walking in a straight line to a tree. He wants to know how many ants are in the line. Zak finds that the line of ants is 40 centimeters long. He finds that the distance between the heads of the ants is 16 millimeters. So he decides that there are 5 ants in 80 millimeters. About how many ants are in the line of ants?

Number of Ants in Line	5	10			
Length of Line (millimeters)	80				

Answer ______________________

Explain Could you use multiples of 10 for the number of ants to make a pattern?

____

____

Ask Yourself

What do I need to do after I find the numbers in the pattern?

Hint Do not forget to add the mass of the bag.

⑤ Rosie and 6 of her friends are cleaning up a trail. They pick up plastic bottles to recycle. Rosie brings back 3 bottles. The next person back brings 6 bottles. The next person brings back 9 bottles. This pattern continues. They put all of the bottles in a bag. Suppose each bottle has a mass of 1 gram. The bag alone has a mass of 5 grams. What is the mass of the filled bag?

Show the pattern in the number of bottles that the friends collected.
3, 6, 9, ________, ________, ________, ________

Answer ______________________

Examine Sophie thinks that the mass of the filled bag is 26 grams. What mistake did Sophie make?

____

____

On Your Own

Solve the problems. Show your work.

6. Evan goes hiking on a trail. Posts are located every 800 meters after the beginning of the trail. The trail measures 16 kilometers. How many posts does Evan pass if he begins at the start of the trail and he stops at 4 kilometers?

Answer ______________________________

Interpret What must you do with the two different units of length to solve the problem?

____

____

7. Dee found some inchworms in her backyard. She noticed that some of the worms are 25 millimeters long and some of them are 50 millimeters long. Suppose she lined up 8 worms that alternated in length: 25 millimeters, 50 millimeters, 25 millimeters, 50 millimeters, and so on. How many centimeters long would the line be?

Answer ______________________________

Propose What is another question the problem could have asked?

____

____

Create

Write and solve a problem about the number of people who hike in a park on each day of the week. Be sure your problem can be solved using the strategy *Look for a Pattern*.

Strategy Focus

Write an Equation

MATH FOCUS: Perimeter and Area

Learn About It

Read the Problem

Linc is fencing off a rectangular play space for his puppy. He wants the play area to be 45 square feet. The play space will be 5 feet wide. What is the total length of fence that Linc needs?

Reread Identify key ideas from the problem.

- What is this problem about?

____

- What does the problem ask you to find?

____

Mark the Text

Search for Information

Reread the problem and circle any measurements that may help you solve it. Keep track of which measurements you know and which ones you must find.

Record For each piece of information, write "Know it" and then record what you know, or write "Do not know it."

- Shape of the play space ____________________
- Area of the play space ____________________
- Length of the play space ____________________
- Width of the play space ____________________
- Length of the fence around the space ____________________

You use the length and width of a rectangle to find its area and its perimeter. Knowing how these measurements are related will help you solve this problem.

Decide What to Do

You know the shape, the area, and the width of the play space. You can use what you know to find what you do not know.

Ask How can I find the total length of the fence?

- I can use the strategy *Write an Equation* to connect what I know to what I do not know.
- I can use the area and width of the play space to find its length. Then I can use the length and width to find the perimeter.

Use Your Ideas

Step 1 Write an equation to find the length of the play space. Use *Area* = *length* × *width*, which is abbreviated $A = lw$. You know two of those measurements. Substitute what you know into the equation.

$\_\_\_\_\_\_\_\_ = l \times \_\_\_\_\_\_\_\_$

What number, when multiplied by 5, gives you 45? ______

So $l =$ ________ feet.

Remember, *A* stands for area, *l* stands for length, *w* stands for width, and *P* stands for perimeter.

Step 2 Write an equation to find the perimeter of the play space. Use $P = 2l + 2w$. You know the length and width. Substitute the measurements of the length and width into the equation.

$P = (2 \times \_\_\_\_\_\_\_\_) + (2 \times \_\_\_\_\_\_\_\_)$

$P = 18 + 10$

$P = 28$ feet

So the perimeter of the play space is ________ feet.

Linc needs ________ feet of fence to go around the play space.

Review Your Work

Look back at the problem. Make sure the measurements given in the problem work with the measurements you found while solving it.

Recognize How does writing an equation help you solve this problem?

____

____

Try It

Solve the problem.

Ask Yourself

How can I use the diagram to figure out the width of the rectangle?

Mark the Text

1. Kendi is planting a vegetable garden in her yard. The diagram shows the shape of her garden. The square measures 2 feet on each side. The rectangles are each 4 feet long. What is the area of the part of the garden that will be planted with corn?

Squash | Lettuce

Corn

Read the Problem and Search for Information

Reread the problem. Find the question you need to answer.

Decide What to Do and Use Your Ideas

You can use the strategy *Write an Equation.*

Step 1 Write an equation that relates what you know to what you need to find out.

- The problem is asking for the ______________ of the corn garden.
- The equation I can write is ______________ .

Step 2 Substitute what you know. Then solve.

- The length of the corn garden is ______________ .
- The width of the corn garden is ______________ .

So the ______________ of the corn garden is ______________ .

Review Your Work

Reread the problem and check that your answer makes sense.

Conclude Suppose the problem was only about the squash garden and the corn garden. Could you still solve the problem? Explain.

____

____

Apply Your Skills

Solve the problems.

(2) Perry has a rectangular pool in the middle of his garden. A fence goes all the way around the edge of the pool to keep everyone safe. The fence is 24 feet around. The pool is 4 feet wide. How long is the pool?

Hint You can draw a diagram to help you.

Is this problem about area or about perimeter? ____________

Substitute the values you know into an equation.

____

Ask Yourself

Which equation should I use?

Solve your equation to find the length of the pool.

Answer ____

Describe How does drawing a diagram help you solve the problem?

____

____

(3) Bob wants to plant tulip bulbs in 2 rectangular flowerbeds. Each of the rectangles measures 8 feet by 3 feet. Bob wants to plant 9 bulbs in each square foot of the beds. How many bulbs does he need?

Hint Notice that there are 2 rectangular flowerbeds.

Is this problem about area or about perimeter? ____________

How many square feet are in Bob's flowerbeds? ________

Ask Yourself

Which clues in the problem can I use to write an equation?

Multiply to find the number of bulbs that Bob needs.

Answer ____

Distinguish What words in the problem helped you decide which equation to write?

____

____

Hint Remember that in a square, length and width are equal.

Ask Yourself

How does the statue affect the space Anya has for her garden?

④ Anya is planting a garden in a park near her house. The garden will form a big square around a statue with a square base. Each side of the outer edge of the garden is 12 feet long. The base of the statue is 4 feet across. How many square feet will Anya have for her garden?

Garden

Statue

Write the equation you will use. ____________

Outer square: ____________________

Base of the statue: ____________________

Answer ______________________________

Identify What steps did you take to solve this problem?

____

____

Hint Remember there are 12 inches in 1 foot.

Ask Yourself

Dora will plant the cactus plants along the edges of each square. What does this detail tell me about the equation I will write?

⑤ Dora works at the Two-Square Desert Ranch. She is planting 2 square gardens in a special way. Here is a diagram of Dora's gardens. Each square measures 20 feet on a side. If Dora plants a cactus every 6 inches along the edge of each square, how many plants will she need?

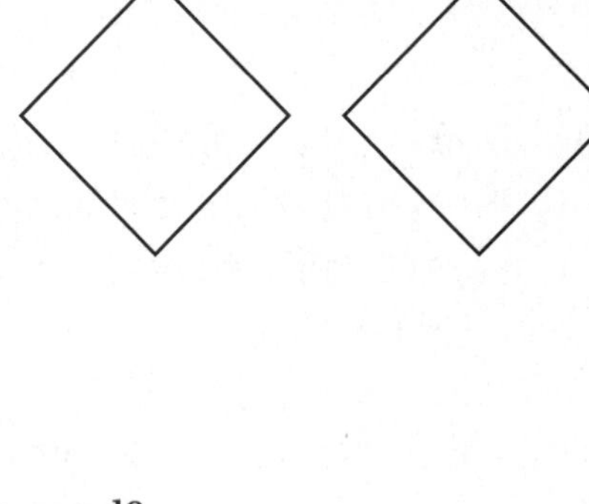

Find the distance around 1 square.

____

Answer ______________________________

Explain Suppose Dora plants a cactus every foot along the edge of each square. How would you solve the problem?

____

____

On Your Own

Solve the problems. Show your work.

⑥ Walter walks his dog along the edge of a rectangular park. The park is 9 feet wide. If he walks his dog once around the park, Walter walks a total distance of 68 feet. What is the area of the park?

Answer ____________________

Determine What is another question this problem could ask?

____

____

⑦ Mia is making a rectangular garden. It will measure 5 feet by 3 feet. She will put bricks all around the outside of the edge of the garden. The bricks are 4 inches long. Mia has 4 square white tiles that she will use at each corner of the garden. How many whole bricks does Mia need?

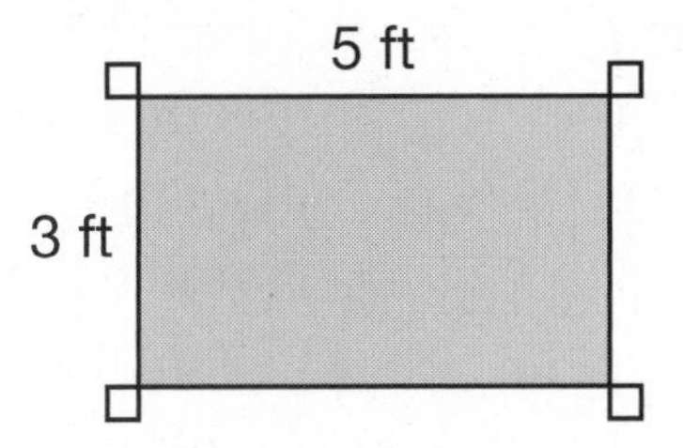

Answer ____________________

Revise Describe the method you used to solve the problem. Explain how you could have solved it in a different way.

____

____

Write and solve a problem about the area of a small dog park completely surrounded by fencing. Be sure your problem can be solved using the strategy *Write an Equation.*

UNIT 5

Review What You Learned

In this unit, you worked with four problem-solving strategies. You can often use more than one strategy to solve a problem. So if a strategy does not seem to be working, try a different one.

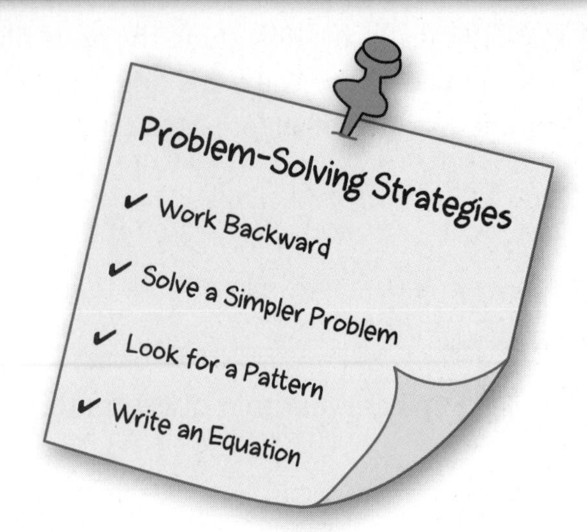

Solve each problem. Show your work. Record the strategy you use.

1. Dr. Sapp experiments with fast-growing vines. Her latest vine triples its length every week. On Monday, April 9, the vine measured 5 millimeters. Dr. Sapp measures the vine every Monday. On what date will the vine measure more than 1 meter long?

Answer ______________________________

Strategy ______________________________

2. A cook weighs a pear, a pineapple, a watermelon, and a grapefruit. The grapefruit and the pear weigh the same. The pear and the grapefruit together weigh half as much as the pineapple. The pineapple weighs twice as much as the pear and the grapefruit together. The watermelon weighs 7 pounds more than the pineapple. How much does the pear weigh?

Answer ______________________________

Strategy ______________________________

3. Goldilocks made some porridge. It was too hot! She waited and measured the temperature again. It was 7°C cooler, but still too hot. So she went for a walk in the forest. When she came back, the temperature had dropped another 24°C. She decided to warm it up again until it was 6°C warmer. Then the temperature was 73°C and it was just right. What temperature was the porridge at the beginning?

Answer ______________________________

Strategy ______________________________

4. Billy is making a rectangular pen for his pet pig. The pen will have an area of 1,500 square feet. It will be 30 feet wide. How many feet of fencing will Billy need to make the pig pen?

Answer ______________________________

Strategy ______________________________

5. Milo has 20 oranges. He wants to use the oranges to outline a square. He will fill the square with other kinds of fruit to make a display for the cafeteria.

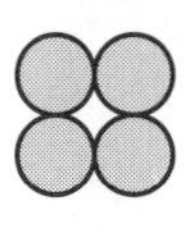
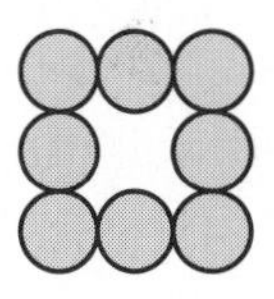
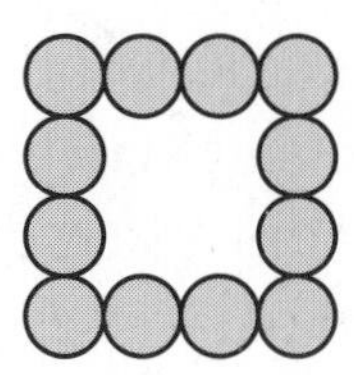

Milo wants to make the largest square he can. How many oranges will he put along each side of his square?

Answer ______________________________

Strategy ______________________________

Explain how you found the size of the square once you knew the number of oranges.

Solve each problem. Show your work. Record the strategy you use.

6. Kayla took part in a walk for charity. She jogged half the total distance of the course. Then she walked with her friends for half the distance that was left. She looked at the time and decided to try to finish quickly. She ran 1,250 meters to the finish line. What was the length, in kilometers, of the whole course?

Answer ______________________________

Strategy ______________________________

7. There are 3 feet in 1 yard. Jen has 72 feet of fencing. She wants to build a square border for her tropical garden. She plans to use all of the fencing to make it. What is the area, in square yards, of the border that Jen plans to build?

Answer ______________________________

Strategy ______________________________

8. Marc is making fruit punch for a party. His recipe makes 1 gallon. The first step says to pour 1 pint of orange juice into a jug. Marc needs to make 6 gallons of punch. How many 2-quart containers of orange juice should he buy?

Answer ______________________________

Strategy ______________________________

Explain the steps you used to solve the problem.

9. Dana has 8 glasses and a pitcher with 2 liters of water. She pours 25 milliliters into the first glass. Then she pours 50 milliliters into the second glass, 75 milliliters into the third glass, and so on. When she has poured 200 milliliters into the eighth glass, how much water is left in the pitcher?

Answer ______________________

Strategy ______________________

10. Rex has a rectangular sign. It is 125 centimeters long. The sign is five times as long as it is wide. What is the perimeter of the sign?

Answer ______________________

Strategy ______________________

Write About It

Look back at Problem 7 and explain the reasoning you used to solve the problem.

____

____

Work Together: Make a Schedule

You are going to take a trip on a plane! Your flight leaves tomorrow at 10:45 A.M. You want to arrive at the airport 2 hours before the flight leaves. You are getting ready to set your alarm clock. What time will you need to wake up?

Plan
1. Work with your team to list all the things you will need to do in the morning. You will have to dress and get to the airport. You may need to finish packing and walk the dog. Remember to eat breakfast!
2. Order the items on the list in any way that makes sense.
3. Estimate the time you will need for each task.

Collect Use your list to make a schedule. Figure out at what time you will finish each task. The first task should be, "Wake up."

Evaluate Look at the schedule. If you wake up at the given time, and do all the tasks, will you get to the airport on time?

Present Show your team's schedule to the class. Discuss how you decided what tasks to put on your schedule and how you made the time estimates.

UNIT 6 Problem Solving Using Data, Graphing, and Probability

Unit Theme:

Exploring

We explore space to learn about other planets. We use technology to understand the weather. We invent robots to help us do things. When people explore, they learn and do many exciting things. In this unit, you will see how math is important to the world of exploration.

Math to Know

In this unit, you will use these math skills:

- Calculate mean, median, mode, and range
- Make and use graphs
- Find combinations and probabilities

Problem-Solving Strategies

- Make an Organized List
- Make a Graph
- Guess, Check, and Revise

Link to the Theme

Finish the story. Write about a robot called Fixit who repairs broken machines for the six families on Spring Street.

Fixit repairs all kinds of gadgets. The families on Spring Street call for his help all the time. By the end of the day, Fixit will visit all six houses on the street.

____

____

____

____

____

____

____

Use Math Language

Review Vocabulary

The list below shows vocabulary terms in this unit. Knowing the meaning of these terms will help you understand the problems.

bar graph	line graph	median	probability
circle graph	mean	mode	range

Vocabulary Activity **Modifiers**

A descriptive word placed in front of another word indicates a specific meaning. Use words from the list above to complete the following sentences.

1. A ______________________ shows changes over time.

2. A ______________________ shows parts of a whole.

3. A ______________________ compares groups of data.

Graphic Organizer **Word Circle**

Complete the graphic organizer.

- Cross out the word that does not belong.
- Replace it with a word from the vocabulary list that does belong.

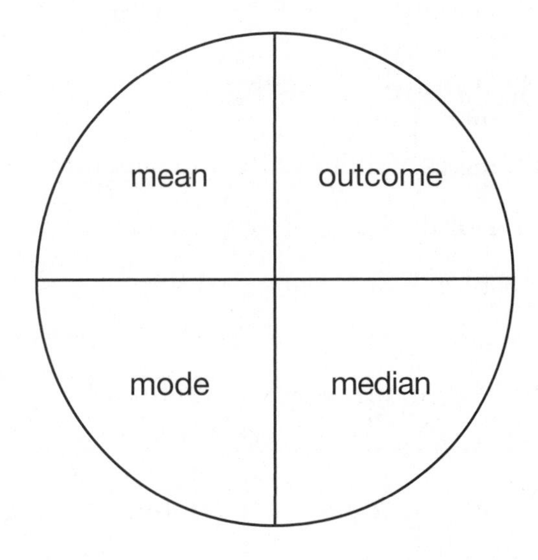

Strategy Focus

Make an Organized List

MATH FOCUS: Mean, Median, Mode, and Range

Learn About It

Read the Problem

The table shows the number of tornadoes in Oklahoma for 5 years. Suppose you are a climate scientist writing a report about this data. You know the mean of the data is 42.8 and the mode is 27. Now you need to find the median and range. What are those two numbers?

Tornadoes in Oklahoma

2005	27
2006	27
2007	49
2008	77
2009	34

Reread Use your own words and ideas to answer the questions.

- What is the problem about?

____

- What information do you have?

____

- What do you need to find?

____

Search for Information

Look at the problem and the table again.

Record How is the table organized?

The years of the tornadoes are in the ______________________ column.

The numbers of tornadoes are in the ______________________ column.

The information you need to choose a strategy is in the problem and in the table.

Decide What to Do

You know the number of tornadoes from 2005 through 2009. You know the mean and the mode of the data. You know the problem asks for two numbers: *median* and *range.*

Ask How can I find the median and the range of the data?

- The median is the middle number when the data are arranged in order from least to greatest. The range is the difference between the greatest number and the least number.
- If I put the numbers in order, it makes it easier to find both the median and the range. So I can use the strategy *Make an Organized List.*

Use Your Ideas

Step 1 List the number of tornadoes in order from least to greatest.

27 27 ______ ______ ______

Least **Greatest**

Check that you included all the data from the table. Also check that the numbers are in order from *least* to *greatest.*

Use the data in the second column of the table on page 174. Making an organized list of all the numbers will help you answer the questions.

Step 2 Find the median in the ordered list above. Circle it. The median is ______ tornadoes.

Step 3 Subtract to find the range.

______ – ______ = ______

greatest number – least number = range

The range is ______ tornadoes.

Review Your Work

Check that you answered both parts of the question.

Explain If the list was ordered from greatest to least, would your answers change? Why or why not?

____

____

Try It

Solve the problem.

① A TV weather scientist displays a line plot showing the number of hurricanes over the Atlantic Ocean for the years 1995–2005.

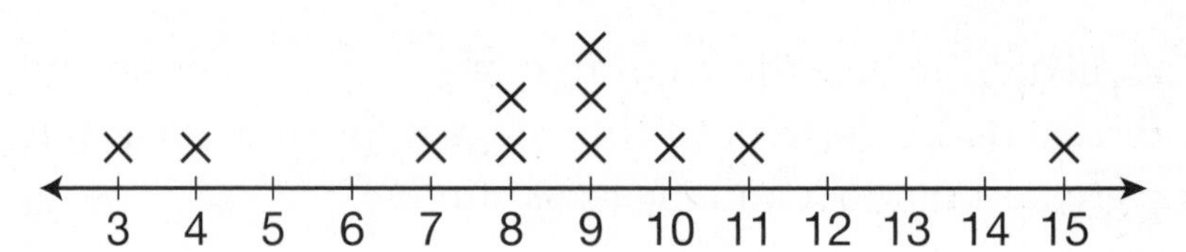

Number of Hurricanes in the Atlantic Ocean (1995–2005)

What are the range and the median for the data?

Mark the Text

Read the Problem and Search for Information

Think about how many years there are from 1995 through 2005.

Decide What to Do and Use Your Ideas

You need to find the range and median of the data for the Atlantic Ocean hurricanes. You can use the line plot to make an organized list. The Xs tell how many times a number should be in the list.

How many Xs are there? ________

Step 1 Find the range. Subtract the least number from the greatest number.

The range is ________ hurricanes.

Step 2 Find the median. Start at the left. Write the number for each X in order. The median is the middle number.

Ask Yourself

What does each X in the line plot mean?

3, 4, 7, 8, 8, 9, ________, ________, ________, ________, ________

The median is ________ hurricanes.

Review Your Work

Check that you listed a number for each X.

Describe How does the line plot show the range and the median?

____

____

Apply Your Skills

Solve the problems.

② Ben lives on the west coast of the United States. His cousin Bev lives on the east coast. During one week in June, the cousins recorded the high temperatures for their cities. Their results are shown in the table below. Which cousin's city had the greater median high temperature that week?

High Temperature of Two Cities (°F)

	Sun.	Mon.	Tue.	Wed.	Thu.	Fri.	Sat.
Ben	91	85	83	75	92	94	80
Bev	79	82	84	89	87	93	92

Hint You need to find the median for *two* different data sets.

Organize the temperatures for each city from least to greatest. Then find the median for each city.

Ben: ______________________ Median: ________

Bev: ______________________ Median: ________

Answer ______________________________

Ask Yourself

What do I have to do after I find the two medians?

Generalize How does writing the numbers in order help you?

____

____

③ Stan recorded how many inches of snow fell every day for a week. His data is listed below. Stan thinks that both the median and the mean are the same for this data. Is he correct? Explain how you know.

Hint To find the mean, divide the total snowfall by the total number of days.

Snowfall (in.)						
5	2	0	8	10	3	7

Ask Yourself

Do I include the zero in my organized list?

Inches of snow: ______________________________
Least **Greatest**

Answer ______________________________

Compare How is this problem like the Learn About It problem? How is it different?

____

____

4 In the town where Rajani lives, a thunderstorm knocked down electrical lines. Many of Rajani's friends had no electricity. Rajani made a line plot of how many hours 14 of her friends had no power. What is the median of the data set?

Hint Note that there is an even number of Xs. When this happens, the median is the mean of the middle two numbers.

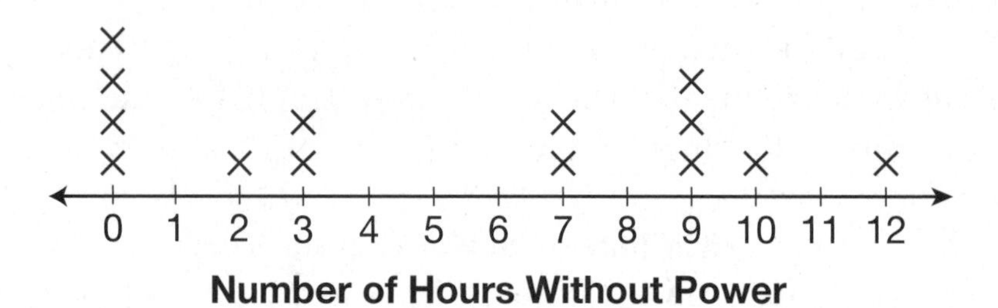

Ask Yourself

In my ordered list, do I need to write a number for each X?

List the numbers in order from least to greatest.

____

Answer ______________________________

Determine Clara found the median to be 8 hours. What mistake did she likely make?

____

____

5 It can be very warm in El Paso, Texas. The table below shows the average number of days each month that the temperature was at least 90°F. Which is greatest—the mean, the median, or the mode? Explain how you know.

Hint To find the mean, divide the total number of days above 90°F by the number of months.

Average Number of Days Each Month Above 90°F in El Paso, Texas

Jan.	Feb.	Mar.	Apr.	May	June	July	Aug.	Sept.	Oct.	Nov.	Dec.
0	0	0	2	14	26	27	24	13	2	0	0

List the numbers in order from least to greatest.

____

Ask Yourself

Can I predict from my organized list if the mean, the median, or the mode will be greatest?

Answer ______________________________

Conclude Which word in the problem tells you that you have to compare numbers to answer the question? How many numbers did you compare?

____

____

On Your Own

Solve the problems. Show your work.

6 The table shows the average number of sunny days in one year in some southern United States cities. Which cities' averages are closest to the median?

Sunny Days in Southern U.S. Cities

City	Average Number of Sunny Days
San Diego, CA	146
Tucson, AZ	193
Roswell, NM	168
San Antonio, TX	105
New Orleans, LA	101
Jackson, MS	111
Mobile, AL	102
Jacksonville, FL	94

Answer ______________________________

Sequence Tell what steps you used to answer the question.

____

____

7 Theo lives in Illinois. His cousin, Teresa, lives in Texas. For 6 weeks, they keep track of how many days it rains in their towns each week.

Number of rainy days in Theo's town: 2, 5, 4, 0, 5, 2
Number of rainy days in Teresa's town: 3, 1, 3, 2, 1, 2
Find the median number of rainy days in each town. Whose town has a greater median number of rainy days? Explain how you know.

Answer ______________________________

Predict Without computing, predict in whose town the mean of rainy days is greater. Explain your prediction.

____

____

Look back at the problems in this lesson for ideas. Write and solve your own problem about median, mode, or range. Be sure your problem can be solved using the strategy *Make an Organized List*.

Strategy Focus

Make a Graph

MATH FOCUS: Bar Graphs, Line Plots, and Line Graphs

Learn About It

Read the Problem

Spirit is a land rover that explores Mars. *Spirit* moves around gathering data about the rocks and soil of Mars. It does not always move at the same speed. The table shows how far *Spirit* moved after different amounts of time passed. About how far had *Spirit* moved after 7 seconds?

Time (seconds)	0	2	4	6	8	10
Distance (centimeters)	0	2	4	6	10	14

Reread Ask yourself these questions as you read.

- What is the problem about?

____

- What data are in the table?

____

- What am I asked to do?

____

Search for Information

Read the problem again. Study the table.

Record What information in the problem and table will help you solve the problem?

Spirit does not always move at the same ________________ .

You need to estimate where *Spirit* was after ________ seconds.

Look at the table. Circle the distances that show where *Spirit* was just before and just after 7 seconds.

Think about how you can picture the data.

Decide What to Do

The information in the table tells you where *Spirit* was after different amounts of time had passed. You need to find where *Spirit* was after a time that is not listed in the table.

Ask How can I find about how far *Spirit* had moved after 7 seconds?

- I can use the strategy *Make a Graph*. A line graph shows how data change over time.
- I can graph the data from the table. Then I can read the graph.

A graph can help you picture the data.

Use Your Ideas

Step 1 Label the horizontal axis with the number of seconds. Use the vertical axis to show the distance *Spirit* had moved compared to where it started.

Label the scale on the vertical axis. Since the distances are all even numbers, a scale of 2 is a good choice.

Step 2 Plot the data from the table on the graph. Connect the points to make the line graph.

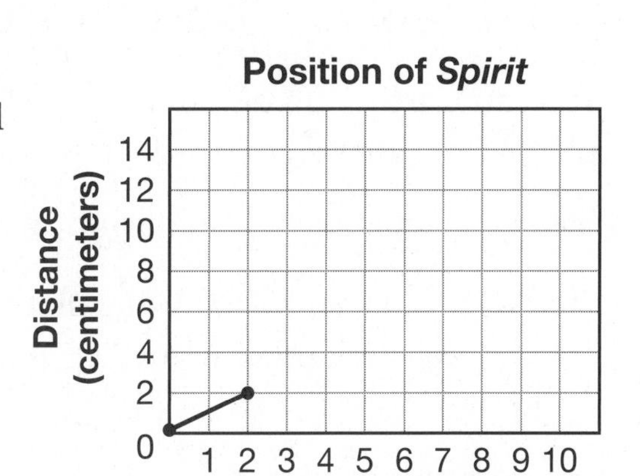

Step 3 Find 7 seconds on the horizontal axis. Next, move up to the line. Then move left to the vertical axis to find the distance at 7 seconds.

After 7 seconds, *Spirit* had moved about ________ centimeters from where it started.

Review Your Work

Check that you plotted the points on the graph correctly.

Describe How does making a graph help you solve the problem?

____

____

Try It

Solve the problem.

① A space shuttle mission takes people and equipment into orbit around Earth. From 1981 through 2002, the shuttles went on 112 missions. The table below shows the number of shuttle missions made each year for these 22 years. What number of missions occurred most often?

Number of Shuttle Missions (1981–2002)
2 3 4 5 9 2 0 2 5 6 6 8 7 7 7 7 8 5 3 5 6 5

Read the Problem and Search for Information

Look at the table to find the numbers you need to use.

Decide What to Do and Use Your Ideas

You can use the strategy *Make a Graph*. Use the data to make a line plot. A line plot helps you see how data are grouped.

Step 1 Label the line plot. You can use the title of the table.

Step 2 Make an X on the line plot for each piece of data. Cross off each number in the list as you plot it.

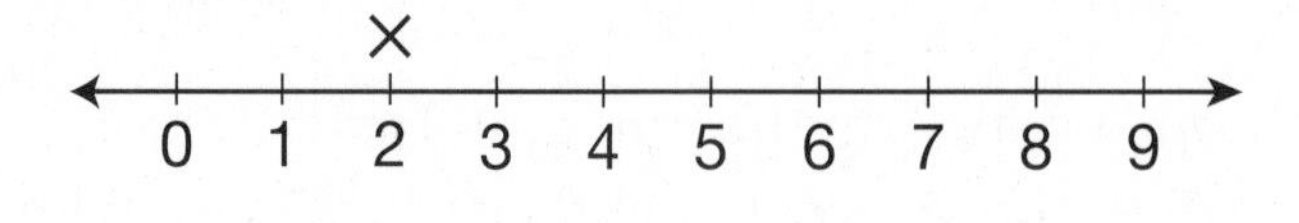

____

Step 3 Use the line plot to find the mode.

The number of missions that occurred most often is ________.

Ask Yourself

How do I find what number occurred most often?

Review Your Work

Check that each piece of data in the table is on the line plot.

Compare How does a line plot make it easier to find the number that occurs most often? Explain.

____

____

Apply Your Skills

Solve the problems.

2 The bar graph shows how many missions 3 space shuttles had from 1981 to 2009. Two other shuttles also had missions. *Discovery* had 37 missions. *Endeavour* had 24 missions. Which shuttle had about $\frac{2}{3}$ as many missions as *Discovery*?

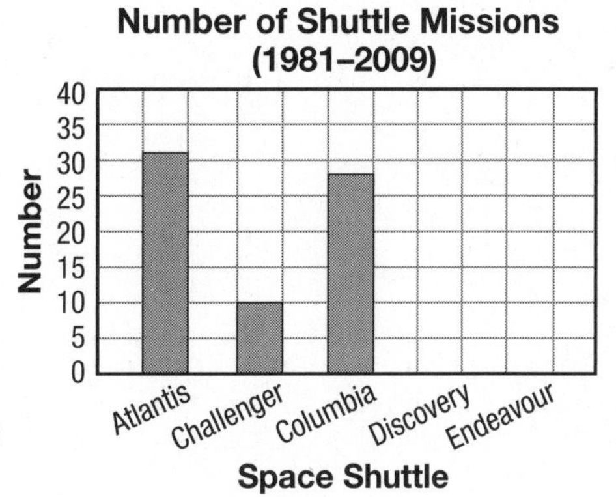

Ask Yourself

How can a bar graph help me answer the question?

Hint Complete the graph to show the data for *Discovery* and *Endeavour*. Divide the bar showing the *Discovery* missions into thirds. Then compare $\frac{2}{3}$ of the *Discovery* bar to the other bars on the graph.

Answer ______________________________

Explain How does a bar graph help you compare data?

____

____

3 The local planetarium has a solar system show every hour. The table shows the number of tickets sold for each show. Between which two shows did ticket sales increase the most?

Planetarium Ticket Sales							
Time	Noon	1:00	2:00	3:00	4:00	5:00	6:00
Tickets	40	50	65	50	70	80	85

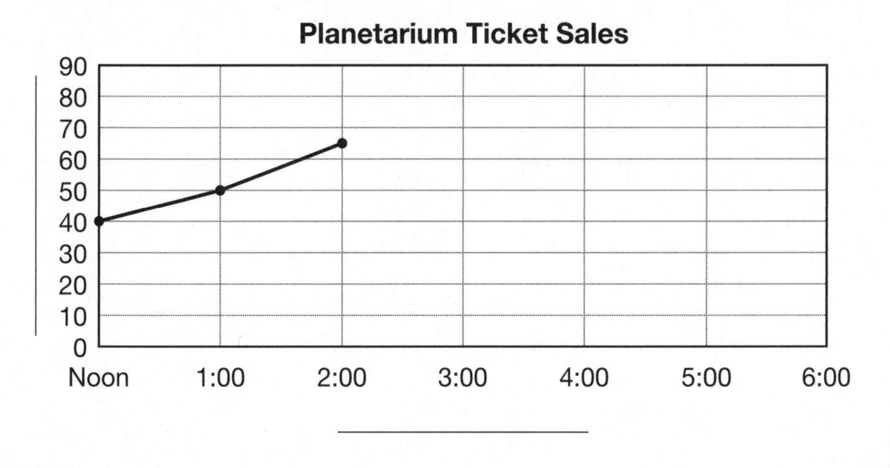

Hint A line graph is a good choice to show changes over time. Complete the graph. Be sure to label the axes.

Ask Yourself

Should I look at where the line rises or where it falls?

Answer ______________________________

State What is another question you can ask given this data?

____

____

Ask Yourself

How can I use a line plot to find the range?

Hint Cross off each number in the list as you make an X above it on the line plot.

4. Ari asks his classmates how many books on space travel each of them has read. His results are shown below. What are the mode and the range of the numbers of books read?

1, 0, 4, 6, 2, 0, 3, 6, 7, 2, 1, 3, 6, 0, 4, 4, 3, 1, 1, 3

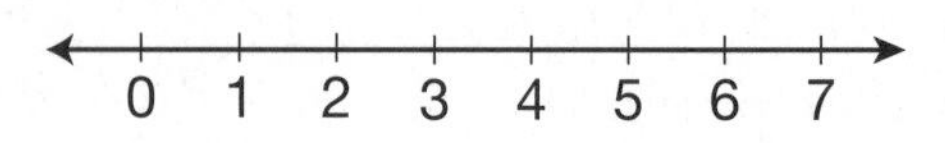

____

Answer ______________________

Determine Ari has read 5 books. How would you change the line plot to show this piece of data?

____

____

Hint Use the information in the problem to complete the graph.

Ask Yourself

How do the heights of the bars for the other moons compare to the bar for Europa?

5. The bar graph shows the number of times the *Galileo* spacecraft flew near three of Jupiter's moons. It also flew near two other moons. *Galileo* flew near Ganymede 6 times and flew near Io 7 times. Which moon did *Galileo* fly near about half the number of times as Europa?

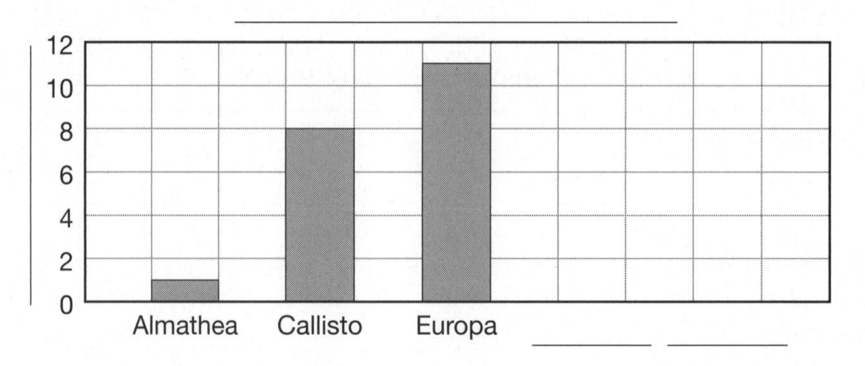

Answer ______________________

Assess How did you use your graph to solve the problem?

____

____

On Your Own

Solve the problems. Show your work.

6. Ms. Myer's class is working on projects about space. The students are working in teams. Which team worked about a quarter of the time of the Satellites?

Time Spent on Space Projects

Time (hours)	Team Name
3	Galaxies
4	Asteroids
6	Planets
8	Comets
11	Satellites

Answer ______________________________

Analyze Suppose you make a bar graph and label and draw the bars in alphabetical order, from Asteroids to Satellites, instead of from least time to most time. Does the order of the bars matter? Explain why or why not.

____

____

7. A crawler carries a space shuttle to the launch pad. Its speed changes often as it moves. The table shows how far the crawler travels over different times. About how far does the crawler travel in 105 minutes?

Distance the Crawler Moves

Time (minutes)	Distance (miles)
0	0
30	0.5
60	1
150	2
210	3

Answer ______________________________

Justify What kind of graph can help you answer this question? Why?

____

____

Create

Write and solve a problem that can be solved by using the strategy *Make a Graph*. Your problem can compare amounts or show change over time.

Lesson 23

Strategy Focus

Guess, Check, and Revise

MATH FOCUS: Data and Circle Graphs

Learn About It

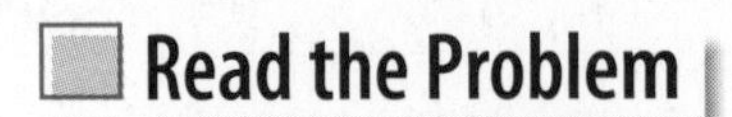

Read the Problem

The botanical garden has plants that catch and eat insects and small animals. There are 72 of these plants on display. There are twice as many sundew plants as Venus flytrap plants. How many more pitcher plants than Venus flytrap plants are on display?

Plants that Eat Insects and Small Animals

Pitcher Plants 33

Sundew Plants

Venus Flytraps

Reread Ask questions as you read.

- What is the problem about?

____

- What does the circle graph show?

____

- What am I trying to find?

____

Search for Information

Read the problem again. Look at the circle graph for information.

Record Write what you know about the number of plants.

The total number of plants is ________ .

The number of pitcher plants is ________ .

There are ____________ as many sundew plants as Venus flytrap plants.

You can use what you know about the number of plants to solve the problem.

Decide What to Do

You know there are 72 plants. There are 33 pitcher plants. So the sum of the number of Venus flytraps and sundews is 72 − 33, or 39. There are twice as many sundews as Venus flytraps.

Ask How can I find how many more pitcher plants than Venus flytrap plants there are?

- I can use the strategy *Guess, Check, and Revise.*
- I can guess the number of Venus flytrap plants. I can check my guess. If my guess is too high or too low, I will revise it.

Use Your Ideas

Step 1 Try 12 for the number of Venus flytraps.

	Flytraps	Sundews	Flytraps + Sundews
Guess 1	12	$2 \times 12 = \square$	$12 + 24 = \square$
Guess 2	15	$2 \times 15 = \square$	$15 + \square = \square$
Guess 3	13	$2 \times 13 = \square$	$13 + \square = \square$

As you work through the steps, complete the table.

Step 2 36 is too low. Try a greater number, like 15, for the number of Venus flytraps.

Step 3 45 is too high. Try a lesser number. When you tried 12, the sum was too low, but closer to 39 than when you tried 15. Try 13.

There are ________ Venus flytraps and ________ pitcher plants in the display. So there are ________ more pitcher plants than Venus flytrap plants.

Review Your Work

Check that you answered the question the problem asked.

Describe Why is *Guess, Check, and Revise* a good strategy to use to solve this problem?

____

____

Try It

Solve the problem.

Animal Eyes

Animal	Class	Number of Eyes
Grasshopper	Insect	5
Scorpion	Arachnid	12
Wolf Spider	Arachnid	8

(1) At the Nature Center, Will saw an exhibit about animal eyes. He saw scorpions and wolf spiders. Altogether, Will saw 7 arachnids and 76 eyes. How many scorpions and wolf spiders did he see?

Read the Problem and Search for Information

Retell the problem in your own words. Mark the information in the table and the problem that will help you answer the question.

Decide What to Do and Use Your Ideas

You can use the strategy *Guess, Check, and Revise.* As you work through the steps, complete the table.

Step 1 You know Will saw a total of 7 arachnids and 76 eyes. Guess the number of scorpions. For your first guess, try 6.

Ask Yourself

Why should I subtract the number of scorpions from 7?

Scorpions	Wolf Spiders	Total Scorpion Eyes	Total Wolf Spider Eyes	Total Arachnid Eyes
6	$7 - 6 = \square$	$6 \times 12 = 72$	$1 \times 8 = \square$	$72 + 8 = \square$
5	$7 - 5 = \square$	$5 \times 12 = \square$	$2 \times 8 = \square$	$\square + \square = \square$

Step 2 80 is too high. Revise your guess. For your second guess, try 5.

So Will saw ________ scorpions and ________ wolf spiders.

Review Your Work

Check that your answer makes sense.

Explain Does it matter if you guess the number of scorpions first or the number of wolf spiders first? Explain your answer.

____

____

Apply Your Skills

Solve the problems.

② In 2010, the total number of endangered frog, toad, and salamander species in the United States was 14. There was 1 more endangered toad species than frog species. There were 3 times as many endangered salamander species as toad species. How many of each species are endangered?

	Number of Frog Species	Number of Toad Species	Number of Salamander Species	Total Number of Endangered Species
Guess 1	3			
Guess 2				
Guess 3				

Ask Yourself How can I use my guess for the number of frog species to find the number of toad species?

Hint If your guess was too high, try a lesser number for the number of frog species.

Answer ______________________

Analyze Why is a table useful when using the strategy *Guess, Check, and Revise?*

____

____

③ Crystal spent 90¢ on 6 stickers. She bought turtle stickers for 25¢ each and bat stickers for 10¢ each. How many more bat stickers than turtle stickers did Crystal buy?

Ask Yourself What number should I use for my first guess?

	Turtle Stickers	Bat Stickers	Cost of Turtle Stickers	Cost of Bat Stickers	Total Cost
Guess 1					
Guess 2					
Guess 3					

Hint The question asks you to compare the number of stickers, not just to find the number of each sticker.

Answer ______________________

Identify What operations did you use to solve the problem?

____

____

Ask Yourself

What label should I use for the last column of the table?

Hint Find the shark that is compared to the other sharks. Start by guessing the length of that shark.

④ At the aquarium, Tim sees 3 different kinds of sharks. The great white shark is twice as long as the bull shark. The bull shark is 2 feet shorter than the tiger shark. The total length of all the sharks is 34 feet. How much longer is the great white shark than the bull shark? Than the tiger shark?

	Length of Tiger Shark	Length of Bull Shark	Length of Great White Shark	
Guess 1				
Guess 2				
Guess 3				

Answer ________________________________

Determine How did knowing that the combined length of the sharks is 34 feet help you?

____

____

Ask Yourself

What numbers and words in the problem and the tally chart will help you solve the problem?

Hint You need to find the total number of emu legs and the total number of llama legs.

⑤ Meg and Tim visit an emu and llama ranch. Emus have 2 legs. Llamas have 4 legs. Meg records the number of animals they see. Tim records the number of animal legs they see. Their results are in the tally chart. How many emus did they see?

Number of Animals	𝍸 II
Number of Legs	𝍸 𝍸 𝍸 III

	Number of Emus	Number of Llamas			
Guess 1					
Guess 2					
Guess 3					

Answer ________________________________

Decide What is another question the problem could have asked?

____

____

On Your Own

Solve the problems. Show your work.

6. A zoologist studies 3 different breeds of large cats. She learns that there were 6 litters of cubs born at a zoo. The table shows how many cubs were in a litter for each of the breeds. There was 1 more litter of jaguar cubs born than cheetah cubs. Altogether, 20 cubs were born. How many litters of lion cubs were born?

Large Cat Litters

Breed	Number in Each Litter
Cheetah	5
Jaguar	2
Lion	4

Answer ______________________________

Evaluate Suppose Bela finds that there were 4 litters of lion cubs born. What error did Bela probably make?

____

____

7. Scientists are studying the eggs of 4 different birds. They have 68 eggs in all to study. There are twice as many owl eggs as hawk eggs. There are 6 more bald eagle eggs than hawk eggs. What is the number of bald eagle eggs? What is the number of hawk eggs?

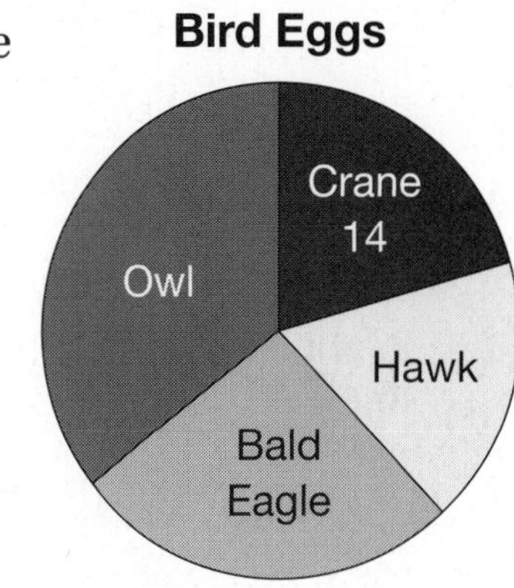

Answer ______________________________

Justify How does the circle graph help you solve the problem?

____

____

Look back at the problems in this lesson. Write and solve your own problem that can be solved by using the strategy *Guess, Check, and Revise.*

Strategy Focus

Make an Organized List

MATH FOCUS: Probability

Learn About It

Read the Problem

A robot in a factory builds toy cars. First the robot chooses a red, green, or blue car body. Next it chooses a red, green, or blue set of wheels. Then the robot puts the wheels on the body to make a car. The choices for body color are equally likely. The choices for wheel color are also equally likely. What is the probability that the color of the car body matches the color of the wheels?

Reread Think of these questions as you reread the problem.

- What is the problem about?

____

- What do I know about the cars?

____

- What do I need to find?

____

Mark the Text

Search for Information

Mark the possible colors for the car body and the possible colors for the set of wheels.

Record Look for details that will help you solve the problem.

How many colors are there for the car body? ________

How many colors are there for the set of wheels? ________

Think about how to use this information to solve the problem.

Decide What to Do

You know what colors can be used for the car parts.

Ask How can I find the probability that the color of the car body matches the color of the wheels?

- I can use the strategy *Make an Organized List.*
- Then I can see how many of the possibilities show matching colors for the body and the wheels.

Use Your Ideas

Step 1 First, the robot chooses a car body. List all the possible cars the robot can put together if it chooses a red body. Next, list all the possible cars the robot can put together if it chooses a green body.

Then list all the possible cars the robot can put together if it chooses a blue body.

Body	Wheels
Red	Red
Red	Green
Red	
Green	Red
Green	
Blue	Red
Blue	

Step 2 Find the probability that the color of the body matches the color of the wheels.

The robot can make ________ different cars.

There are ________ cars that have the same color body and wheels.

________ out of ________ outcomes are favorable.

The probability that the color of the car body matches the color of the wheels is ________ .

The probability is the number of favorable outcomes divided by the number of all possible outcomes. Use your list to count the outcomes.

Review Your Work

Make sure that your list shows all the possible color combinations.

Tell How did making an organized list help you?

____

____

Try It

Solve the problem.

1. Students in a science club are planning a robot contest. Each robot will do a spin, a roll, and a jump. Each robot must do all 3 tasks, one at a time, in a different order from the other robots. In the contest, every possible order of tasks will happen only once. How many robots will be in the contest?

Read the Problem and Search for Information

Reread the problem and use your own words to retell the problem. See what the contest plans tell you about the tasks for the robots.

Decide What to Do and Use Your Ideas

You can use the strategy *Make an Organized List* to order all of the different ways a robot can do a spin, a roll, and a jump. Then you can count the number of combinations on the list.

Ask Yourself

If the first task is a spin, what choices are left for the second task? After you know two tasks, what choice is left for the third one?

Step 1 List all the ways to arrange the tasks if a *spin* is first.
List all the ways to arrange the tasks if a *roll* is first.
List all the ways to arrange the tasks if a *jump* is first.

First Task	Second Task	Third Task
Spin	Roll	Jump
Spin	Jump	Roll
Roll	Spin	
Roll	Jump	
Jump		

Step 2 There are ________ different ways to arrange the tasks.

So there are ________ robots in the competition.

Review Your Work

Check that you did not miss or repeat any sets of tasks.

Recognize Why is *spin, spin, roll* not a possibility on your list?

____

____

Apply Your Skills

Solve the problems.

② Nora is naming her new robot. The name will have one letter and one number. Nora wants the robot's name to start with the letter R or the letter D. She wants the number after the letter to be 2, 3, 4, or 5. If each choice is equally likely, what is the probability that her robot's name will be D2?

Letters R, D	Numbers 2, 3, 4, 5
R	2
R	3
R	4
R	
D	

Hint List all the possible names that start with R. Then list all the names that start with D.

Ask Yourself

How many possible outcomes are there? How many favorable outcomes are there?

Answer ______________________________

Explain Jamie said that the probability Nora's robot will have the name D2 is $\frac{1}{5}$. What error might he have made?

____

____

③ Marco has a robotic dog. It can sit, roll over, bark, and lie down. Marco wants the dog to do 4 different tricks in a row. He does not want the dog to repeat tricks. The dog always barks first. How many ways can the 4 tricks be arranged?

Hint You only need to arrange 3 of the tricks.

First Trick	Second Trick	Third Trick	Fourth Trick
Bark	Sit	Roll Over	Lie Down
Bark	Sit	Lie Down	
Bark	Roll Over	Sit	
Bark	Roll Over		
Bark			
Bark			

Ask Yourself

If the dog sits for the second trick, what are the last two possible tricks that it can do?

Answer ______________________________

Contrast How would your list be different if the dog did not always bark first?

____

____

Ask Yourself

Do I have all the possibilities?

Hint You can use letters to stand for the different types of juice. Use A for apple, B for berry, C for cherry, G for grape, and O for orange.

④ Leslie asks her robot to bring her 2 juice boxes. The robot can choose from apple, berry, cherry, grape, and orange. Without looking, the robot picks 2 juice boxes. What is the probability that the robot will bring Leslie a cherry juice box and a grape juice box?

Pairs with apple juice

A	B
A	

Pairs with berry juice not already listed

B	C

Pairs with cherry juice not already listed

C	G
C	

Pairs with orange juice not already listed

O	

Answer ______________________________

Decide What if the robot had 6 different kinds of juice boxes to choose from? How would that change your list of possibilities?

____

____

Hint There are 3 different heads, 1 kind of body, and 3 different feet that can be used to build the robots.

⑤ A company builds robots. Workers put together different heads, bodies, and feet. Heads can have 1 eye, 2 eyes, or 3 eyes. The body is always a cone. Feet can be blocks, wheels, or boots. How many different robots are there?

Ask Yourself

Is there a simple way to solve this problem?

My list shows that a robot with 1 eye and a cone body can have ________ kinds of feet.

Answer ______________________________

Conclude How is this problem like Problem 3?

____

____

On Your Own

Solve the problems. Show your work.

6. A scientist built a robot that shows its feelings. It took the scientist 5 months to build the robot. The robot shows 3 different robot faces to a group of people. In how many different orders can the scientist show the 3 robot faces?

Answer ______________________________

Discuss What information is not needed to solve the problem?

____

____

7. Derek eats at the Downtown Diner. The diner has a dancing robot. The robot can do a spin, a windmill, and a back flip. Each move is equally likely to be performed. The robot only does 2 different moves in each dance. What is the probability that the robot's first move is a spin?

Answer ______________________________

Modify What is another question the problem could have asked?

____

____

Create

Write and solve your own problem about possible outcomes and probability. Be sure the problem can be solved using the strategy *Make an Organized List.*

UNIT 6 Review What You Learned

In this unit, you worked with three problem-solving strategies. You can often use more than one strategy to solve a problem. So if a strategy does not seem to be working, try a different one.

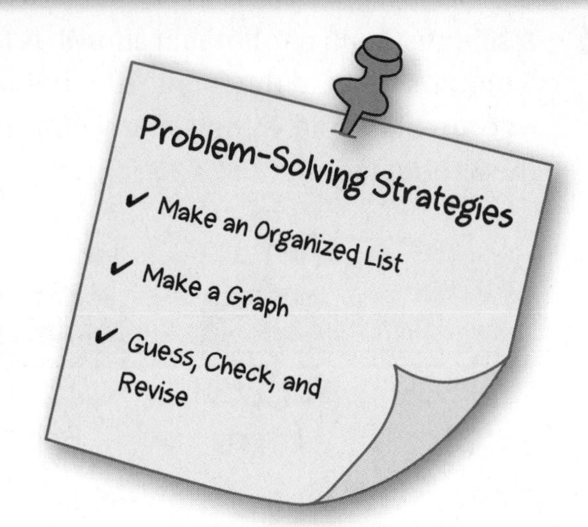

Solve each problem. Show your work. Record the strategy you use.

1. Hector buys a book about fossils. He gets 7 coins totaling $0.42 in change. Does Hector's change include a quarter? Which coins does Hector receive in change?

Answer ______________________________

Strategy ______________________________

2. For 7 days in July, Jason keeps track of the number of moths and fireflies he sees. He recorded the numbers in the table below. What is the median number of moths? Of fireflies?

Number of Insects Each Day

Moths	13	22	11	9	34	20	11
Fireflies	12	15	13	11	12	14	12

Answer ______________________________

Strategy ______________________________

3. Jo plants a sunflower. She measures the height of the plant every week. The table shows the height of the plant each week. About how tall is the sunflower at 20 days?

Sunflower Growth

Day	Height (inches)
0	0
7	7
14	15
21	27
28	40
35	52

Answer ______________________________

Strategy ______________________________

4. Sheila is going to sew these 3 beads onto the front of a belt. In how many different orders can she sew the 3 beads in a row?

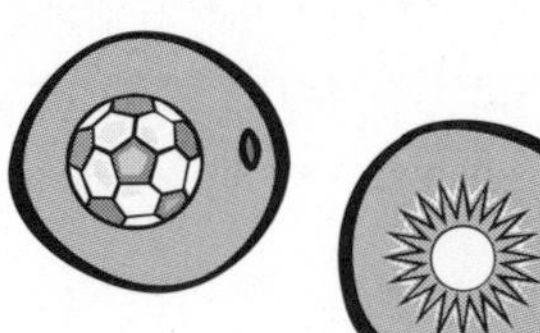
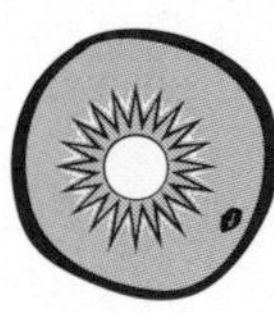

Answer ______________________________

Strategy ______________________________

5. Luis surveyed 36 students about their favorite season. The circle graph shows the results. Half as many students chose winter as summer. Two more students chose spring than fall. How many more students chose spring than winter as their favorite season?

Survey of Favorite Seasons

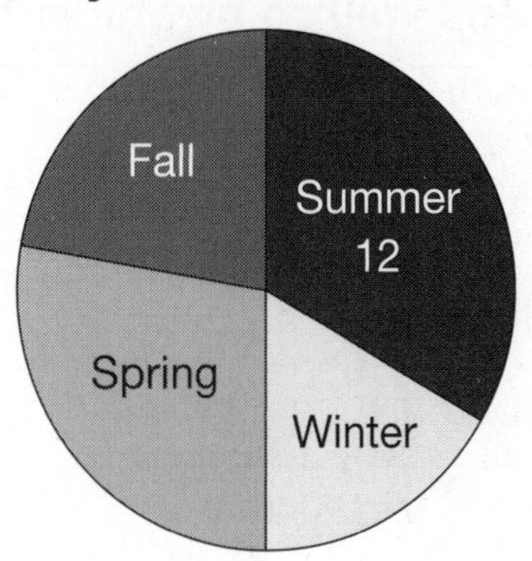

Answer ______________________________

Strategy ______________________________

Explain how you can use the graph to check your answer.

Solve each problem. Show your work. Record the strategy you use.

6. Five students joined a walk-a-thon to raise money for the school library. Anita walked 19 laps. Connor walked 24 laps. The number of laps that Rosa, Max, and Jamal walked are shown on the graph. Which friend walked about one third as many laps as Rosa?

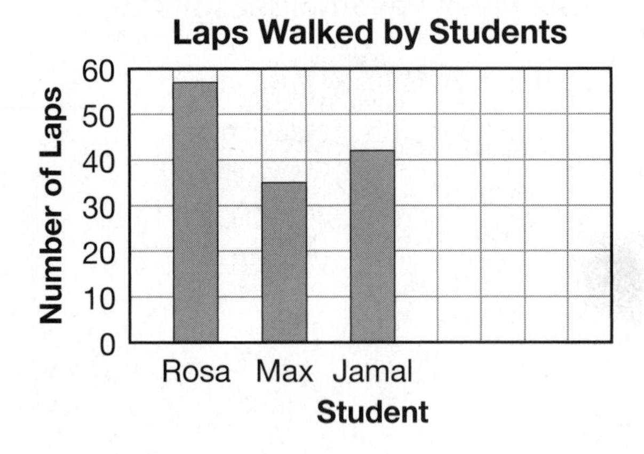

Answer ______________________

Strategy ______________________

7. The table shows the number of hours Jamie volunteers on different days at the animal shelter. What are the mode and range of the number of volunteer hours?

Number of Volunteer Hours												
6	5	8	2	5	4	8	7	1	8	7	5	8

Answer ______________________

Strategy ______________________

8. Mrs. Diaz grows pumpkins to sell at the farmers' market. She records the weights of the first 10 pumpkins she picks. Is the mean greater than or less than the median? Explain how you know.

Pumpkin Weights (pounds)									
12	23	27	7	10	28	30	23	26	14

Answer ______________________

Strategy ______________________

Explain why organizing the data helps you to solve the problem.

9. Marco planted tulip and daffodil bulbs. He used 8 packages of bulbs and planted 20 bulbs. Each package of tulip bulbs had 2 bulbs. Each package of daffodil bulbs had 3 bulbs. How many of each type of bulb did Marco plant?

Answer ______________________

Strategy ______________________

10. The Bears, Eagles, Falcons, Jaguars, and Panthers compete in a tournament. Without looking, Tom will draw the names of the two teams that will play the opening game out of a hat. What is the probability that the Panthers will play the Eagles in the opening game?

Answer ______________________

Strategy ______________________

Write About It

Look back at Problem 1. How did you solve the problem?

____

____

Work Together: Plan an Aquarium

Your team is creating a freshwater aquarium. You have $14 to spend on 1 fish, 1 plant, and 1 decoration for your aquarium. What combination will you choose?

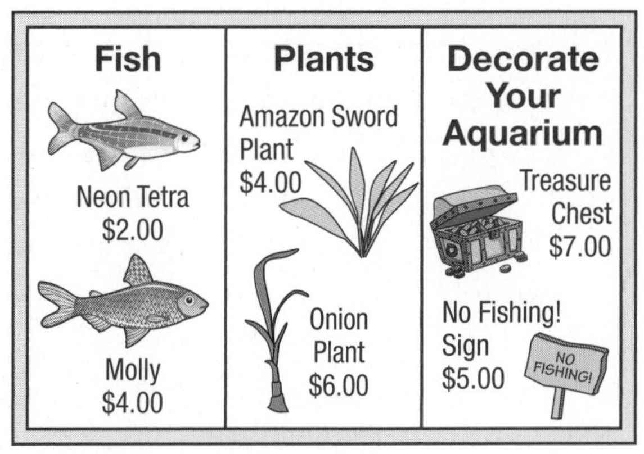

Plan Find the possible combinations and their prices.

Decide As a group, choose the combination of 1 fish, 1 plant, and 1 decoration you will buy.

Justify Explain your team's decision. Use tables, lists, or a bar graph to compare the amounts you will spend on a fish, a plant, and a decoration.

Present As a group, share your decision and bar graph with the class.

Math Vocabulary Activities

On the next six pages are some of the math terms you have worked with in each unit.

You can cut these pages to make vocabulary cards. The games and activities below can help you learn and remember the meaning of these important terms.

Try This!

- Complete the activity on the back of each vocabulary card. Use a separate sheet of paper. Discuss your work with a partner or in a small group.
- Work with a partner. Take turns. One person chooses a vocabulary card and shows the front of the card. The other person gives the definition. Check to see if your partner was correct by looking at the back of the card.
- Draw a picture or write an example of each term on a separate card. Then have a partner match your example or picture to a vocabulary card.
- Play a matching game with a partner or a small group.

  Make your own sets of cards. Make one set for each term. Write a term on one side and leave the other side blank. Lay out the cards in this set in rows, facedown. Make another set for each definition. Write a definition on one side and leave the other side blank. Lay out the cards in this second set in rows, facedown and separate from the first set.

  Take turns. The first player turns over two cards, one from each set. If the cards show a word and its matching definition, the player keeps them and takes another turn. If the word and definition do not match, place the cards facedown where they were, and it is the next player's turn. The player with the most matched pairs wins the game.

Math Vocabulary

Unit 1

difference

15 − 6 = 9
↑
difference

digit

0, 1, 2, 3, 4, 5, 6, 7, 8, 9

equation

10 + 20 = 30

expression

10 + 8

pattern

2, 4, 6, 8, 10, ...

place value

Hundreds	Tens	Ones
2	7	5

rule

5, 10, 15, 20, ...
+ 5
↑
rule

sum

6 + 4 = 10
↑
sum

Unit 2

dividend

60 ÷ 2 = 30
↑
dividend

divisor

60 ÷ 2 = 30
↑
divisor

factor

15 × 2 = 30
↑ ↑
factors

multiple

Multiples of 4:
0, 4, 8, 12, 16, ...

product

15 × 2 = 30
↑
product

quotient

60 ÷ 2 = 30
↑
quotient

remainder

10 ÷ 6 = 1 R4
↑
remainder

Venn diagram

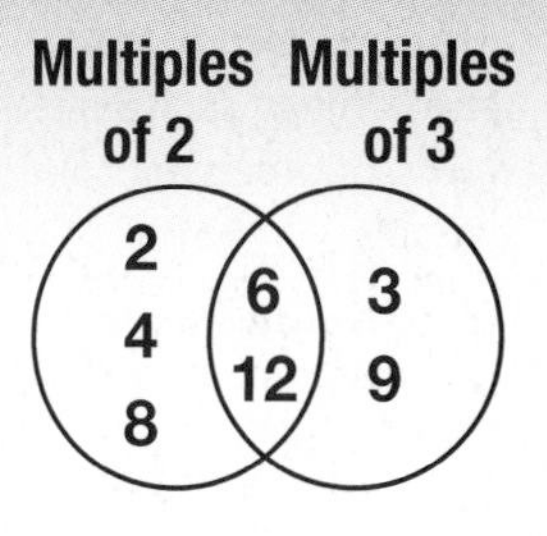

Math Vocabulary

Unit 1

expression

a combination of numbers and at least one operation symbol, but no equals sign

Give an example of the word.

equation

a number sentence that has an equals sign in it

Give an example of the word.

digit

any of the symbols 0, 1, 2, 3, 4, 5, 6, 7, 8, 9

Write a sentence using the word.

difference

the answer in subtraction

Give an example of the word.

sum

the answer in addition

Give an example of the word.

rule

a way to describe how objects or numbers are arranged in a pattern

Write a definition in your own words.

place value

the value of a digit's place in a number

Write a sentence using the word.

pattern

objects or numbers arranged according to a rule or rules

Draw a picture that shows a pattern.

Unit 2

multiple

A multiple of a number is the product of that number and some whole number.

Give examples.

factor

a number that is multiplied by another number to get a product

Give an example of the word.

divisor

a number by which another number is divided

Give an example of the word.

dividend

a number that is being divided

Give an example of the word.

Venn diagram

a diagram used to relate sets of things, such as words, numbers, or symbols

Draw an example of a Venn diagram.

remainder

the number left over after a quotient is found

Give an example of the word.

quotient

the answer in division

Give an example of the word.

product

the answer in multiplication

Give an example of the word.

Math Vocabulary

Unit 3

decimal

decimal point

0.25

tenths

hundredths

decimal point

0.25

decimal point

denominator

$\frac{3}{4}$

denominator

equivalent

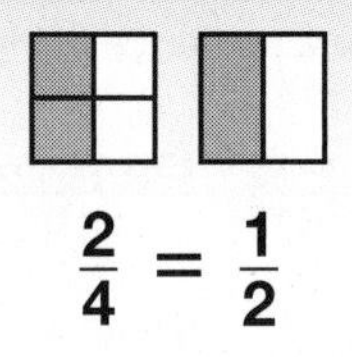

$\frac{2}{4} = \frac{1}{2}$

fraction

$\frac{1}{4}$

hundredths

0.25

hundredths

numerator

numerator

$\frac{3}{4}$

tenths

0.25

tenths

Unit 4

horizontal

prism

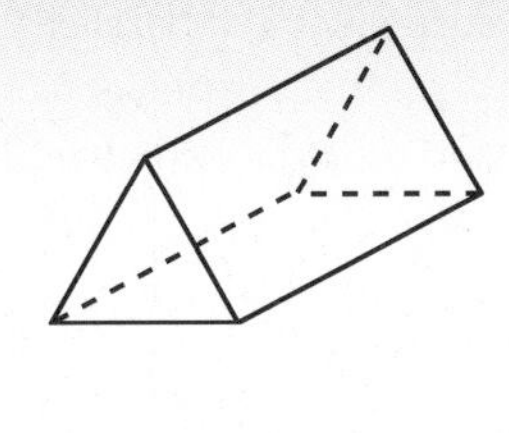

pyramid

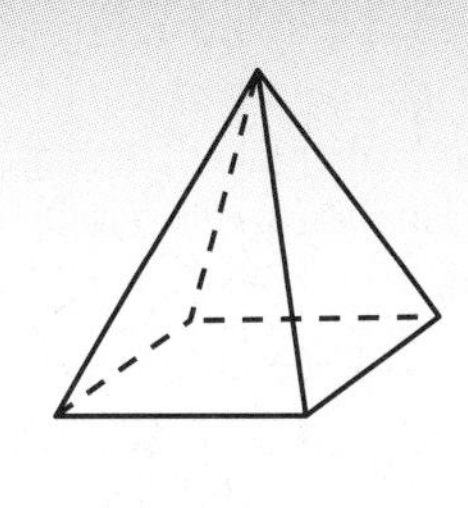

reflection

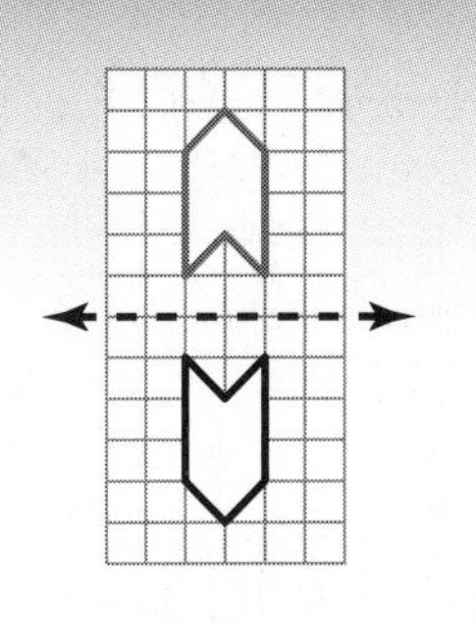

rotation

transformation

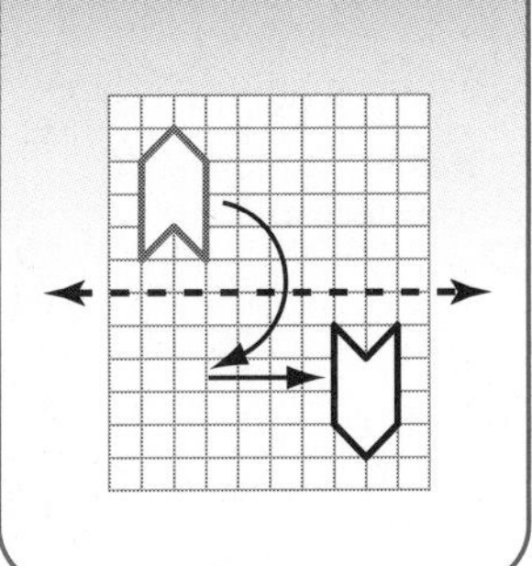

translation

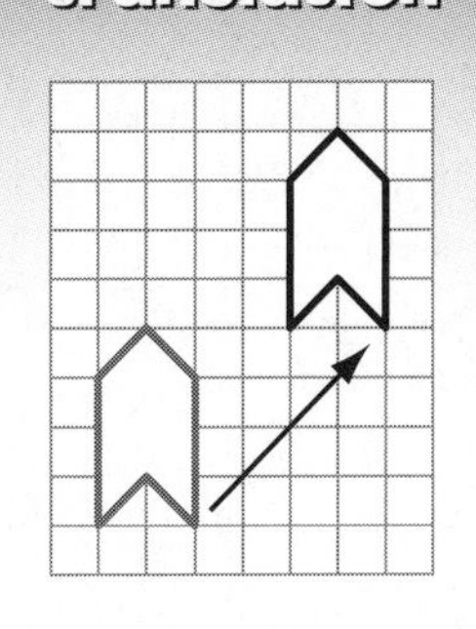

vertical

Math Vocabulary

Unit 3

equivalent

having the same value

Write a sentence using the word.

denominator

the number in a fraction below the bar that tells how many equal parts there are in a whole or a set

Give an example of the word.

decimal point

the symbol in a decimal that separates the whole number from the decimal part

Give an example of the word.

decimal

a number written with a decimal point and digits to the left and right of the point

Write a sentence using the word.

tenths

parts of a whole divided into 10 equal parts

Give an example of the word.

numerator

the number in a fraction above the bar that tells how many equal parts of the whole or the set you are talking about

Give an example of the word.

hundredths

parts of a whole divided into one hundred equal parts

Give an example of the word.

fraction

a number that describes a part of a whole or a set

Give an example of the word.

Unit 4

reflection

a transformation that gives a mirror image of a figure, also called a *flip*

Draw a picture of a reflection.

pyramid

a solid that has a polygon for a base and other faces that are triangles that meet at a point

Draw a pyramid.

prism

a solid with two congruent and parallel bases that are polygons, and other faces that are rectangles

Is a cube a prism? Explain.

horizontal

going left and right

Draw a picture that shows the word.

vertical

going straight up and down

Draw an example of the word.

translation

a transformation of a figure by sliding it, also called a *slide*

Draw an example of this word.

transformation

a movement of a figure

Name 3 examples of a transformation.

rotation

a transformation that turns a figure around a point, also called a *turn*

Draw a picture of a rotation.

Math Vocabulary

Unit 5

area

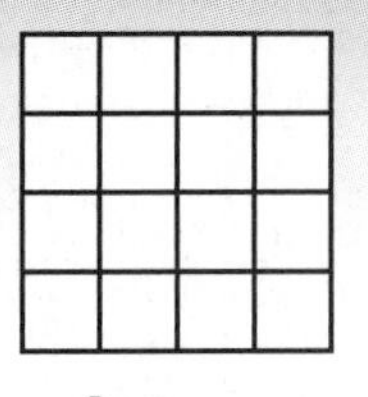

Area =
16 square units

centimeter

1 centimeter

gram

The mass of
a large [paper clip] is
about 1 gram.

kilogram

A [book] weighs
about
1 kilogram.

meter

1 meter

milliliter

A 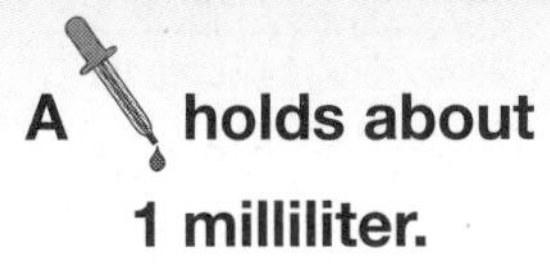 holds about
1 milliliter.

millimeter

1 millimeter

perimeter

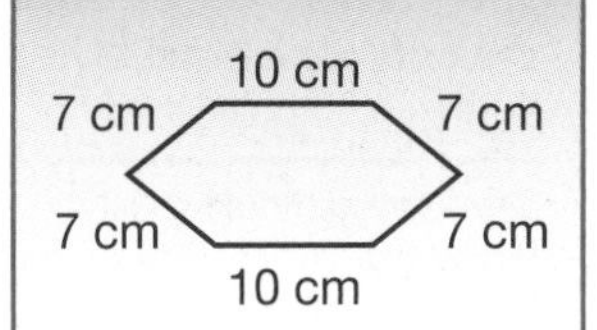

Perimeter =
48 centimeters

Unit 6

bar graph

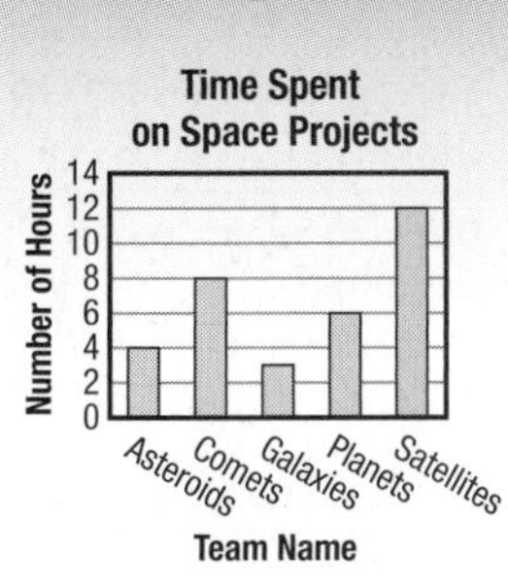

circle graph

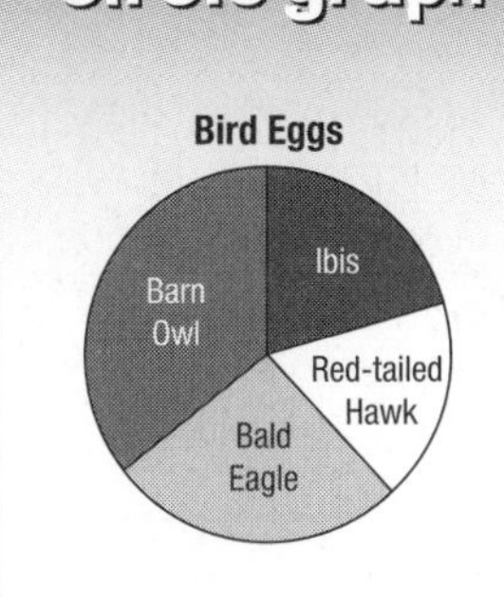

line graph

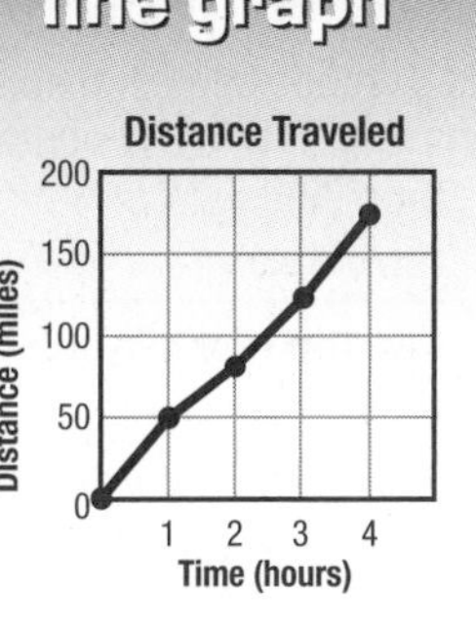

mean

The mean of 7
and 3 is **5**:

$$\frac{7 + 3}{2} = \mathbf{5}$$

median

25
31
78 ← median
84
99

mode

15
13
20
28
7
13

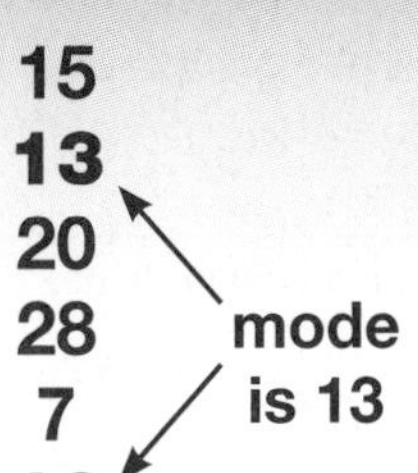

mode
is 13

probability

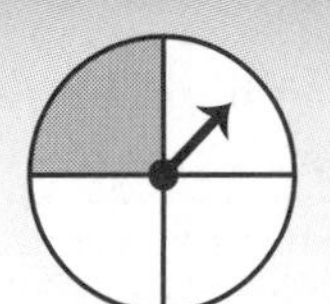

The probability
of spinning white
is $\frac{3}{4}$.

range

34, 35, 36, 37,
38, **39**

39 – 34 = **5**
↑
range

Math Vocabulary

Unit 5

kilogram

a basic metric unit of mass, which is equal to 1,000 grams

Write a sentence using the word.

gram

a metric unit of mass equal to 0.001 kilogram

Use the word in a sentence.

centimeter

a metric unit of length equal to 0.01 meter

Write a sentence using the word.

area

the amount of surface enclosed in a figure, usually measured by the number of square units needed to cover the figure

Use the word in a sentence.

perimeter

the distance around a figure

Draw an example of the word.

millimeter

a metric unit of length equal to 0.001 meter

Write a sentence using the word.

millimeter

a metric unit of capacity equal to 0.001 liter

Write a sentence using of the word.

meter

the basic metric unit of length, which is equal to 1,000 millimeters

Write a definition in your own words.

Unit 6

mean

the quantity found by adding the numbers in a set of data and dividing the sum by the number of numbers in the set

Give an example of the word.

line graph

a graph that uses one or more line segments to show data

Draw an example of the word.

circle graph

a graph in which data are represented by parts of a circle

Draw an example of a circle graph.

bar graph

a graph that uses bars to show data

Draw an example of a bar graph.

range

the difference between the greatest number and the least number in a set of data

Write a sentence using the word.

probability

the chance of an event happening

Use the word in a sentence.

mode

the number that appears most often in a set of data

Give an example of the word.

median

the middle number in a set of data when the numbers are put in order

Give an example of the word.